LIGHT
THE DIVINE INTELLIGENCE

Robert Goodwin

Light : The Divine Intelligence
© 2016 Robert Goodwin.

ISBN 978-0-9572745-7-0

All rights reserved.

No part of this publication may be reproduced or transmitted in any form or by any means, electronic or mechanical, including photocopying, recording or any information storage or retrieval system, without either prior permission in writing from the publisher or a licence permitting restricted copying.

Although the author and publisher have made every effort to ensure that the information in this book was correct at the time of going to press, the author and publisher do not assume and hereby disclaim any liability to any party for any loss, damage, or disruption caused by errors or omissions, whether such errors or omissions result from negligence, accident, or any other cause.

Cover image under licence from https://us.fotolia.com/

Other books by the same author:

Truth from the White Brotherhood (1998)
The Golden Thread (1999, reprinted 2005)
Transcognitive Spirituality (2013, reprinted 2015)

With Co-author Amanda Goodwin

Answers for an Enquiring Mind (2002)
In the Presence of White Feather (2005, reprinted 2010)
The Enlightened Soul (2008)
The Collected Wisdom of White Feather (2010)

Published by R.A. Associates 2016
Designed and printed in the UK
mail@whitefeather.org.uk

"You have to see, not through the eyes of another.
This light, this law, is neither yours nor that of another.
There is only light. This is love."

- J. Krishnamurti (1895-1986)

light1 *noun* **1** a form of electromagnetic radiation that travels freely through space, and can be absorbed and reflected, especially that part of the spectrum which can be seen with the human eye. **2** any source of light, such as the Sun, a lamp, a candle, etc. **3** an appearance of brightness; a shine or gleam • *see a light away in the distance.* **4 (the lights)** traffic lights • *turn left at the lights.* **5** the time during the day when it is daylight. **6** dawn. **7** a particular quality or amount of light • *a good light for taking photographs.* **8** a flame or spark for igniting. **9** a means of producing a flame for igniting, such as a match. **10** a way in which something is thought of or regarded • *see the problem in a new light.* **11** a hint, clue or help towards understanding. **12** a glow in the eyes or on the face as a sign of energy, liveliness, happiness or excitement. **13** someone who is well regarded in a particular field • *a leading light.* **14** an opening in a wall that lets in light, such as a window. **15 (lights)** *formal* someone's mental ability, knowledge or understanding • *act according to one's lights.* *adj* **1** having light; not dark. **2** said of a colour: pale; closer to white than black. *verb (past tense and past participle* **lit** *or* **lighted**, *present participle* **lighting**) **1** to provide light for something • *light the stage.* **2** *tr & intr* to begin to burn, or to make something begin to burn • *light the fire.* **3** to guide or show someone the way using a light or torch. **4** *tr & intr* to make or become bright, sparkling with liveliness, happiness or excitement. **lightish** *adj.* **lightness** *noun.* **bring something to light** to make it known or cause it to be noticed. **come to light** to be made known or discovered. **go out like a light** to fall sound asleep soon after going to bed. **hide one's light under a bushel** see under *bushel.* **in a good** or **bad light** putting a favourable or unfavourable construction on something. **in the light of something** taking it into consideration. **light at the end of the tunnel** an indication of success or completion. **lights out 1** *military* a bugle or trumpet call for lights to be put out. **2** the time at night when lights in a dormitory or barracks have to be put out. **see the light 1** to understand something. **2** to have a religious conversion. **see the light of day 1** to be born, discovered or produced. **2** to come to public notice. **shed** or **throw** light on something to make it clear or help to explain it. **strike a light!** *chiefly Austral slang* expressing surprise.
ETYMOLOGY: Anglo-Saxon *leoht.*
Definitions taken from The Chambers Dictionary (online version)
http://www.chambers.co.uk/dictionaries/the-chambers-dictionary.php

This book is dedicated to my close family, with my boundless love and gratitude.

.....and to thinkers everywhere.

There are two ways of spreading light,
to be the candle or the mirror that reflects it.
Edith Wharton

My position is perfectly definite. Gravitation, motion, heat, light, electricity and chemical action are one and the same object in various forms of manifestation.
Robert Mayer

Light brings us the news of the Universe.
Sir William Bragg

Light gives of itself freely, filling all available space. It does not seek anything in return; it asks not whether you are friend or foe. It gives of itself and is not thereby diminished.
Michael Strassfeld

The changing of bodies into light, and light into bodies, is very conformable to the course of nature, which seems delighted with transmutations.
Sir Isaac Newton

When you possess light within, you see it externally.
Anaïs Nin

Life is composed of lines of light, fibres of energy, pulsing, still, pure being.
Frederick Lenz

Look how the smaller birds greet the sun, with so much merry chirruping and so much outpouring of song! It is their way of expressing worship for the only Light they can know, an outer one. But man can also know the inner Sun, the Light of the Overself. How much more reason has he to chirp and sing than the little birds! Yet how few men feel gratitude for such privilege.
Paul Brunton

According to Vedanta, there are only two symptoms of enlightenment, just two indications that a transformation is taking place within you toward a higher consciousness. The first symptom is that you stop worrying. Things don't bother you anymore. You become light-hearted and full of joy. The second symptom is that you encounter more and more meaningful coincidences in your life, more and more synchronicities. And this accelerates to the point where you actually experience the miraculous.
Deepak Chopra

Within each of us is a light, awake, encoded in the fibres of our existence. Divine ecstasy is the totality of this marvellous creation experienced in the hearts of humanity.
Tony Samra

There is no light at the end of the tunnel. There is only you. You are the light.
Author unknown

Contents

Introduction 11

Part One - The Outer Light

Chapter 1: ... and the universe said 'Let there be light' 17

Chapter 2: Ghost in the Machine 43

Chapter 3: Deeper Fields 71

Part Two - The Inner Light

Chapter 4: The Spectrum of Consciousness 99

Chapter 5: Gifts of the Light 131

Chapter 6: Another Perspective 159

Chapter 7: Making Light Work 185

Chapter 8: Enlighten-*meant* 217

Postscript *227*

Bibliography *229*

Online Sources *230*

Light: The Divine Intelligence

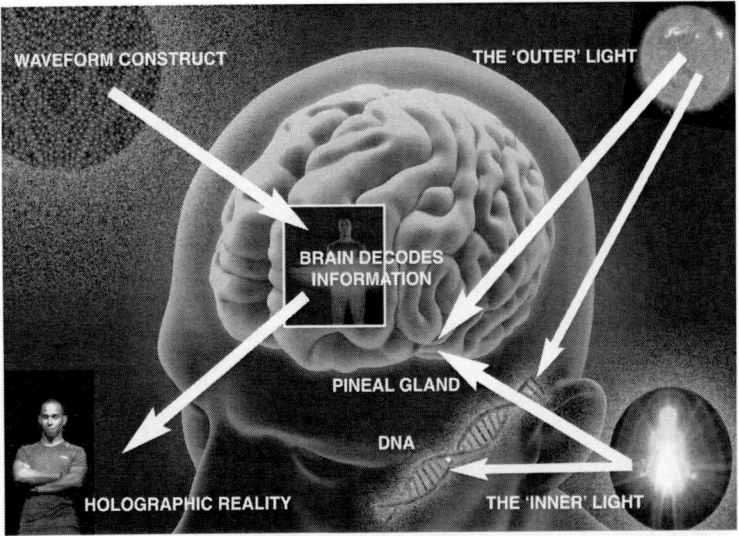

The Map

Our Consciousness decodes energy/information from the Akash (Universal waveform construct) and from other resonant fields with which we are compatible.

Our Consciousness decodes energy/information from the Sun and other luminous bodies within the cosmos. The pineal gland and our DNA are vital aspects of this process.

Our Consciousness decodes energy/information from our own higher self (Soul) and the Inner Light that emerges through our spiritual development.

Our Consciousness, operating through the Mind/Brain interface further decodes energy/information, converting it into the holographic world we experience as solid reality.

Light/Energy/Information from all sources is used to maintain our physical 'body' in a healthy state, repair damage and enable us to perceive dimensions existing beyond the five senses.

All of this, all that we are, is LIGHT.

Introduction

Following the publication of my last book *Transcognitive Spirituality* I felt a sense of deep satisfaction. The manuscript had taken me over twelve months of research and writing to assemble and if you'll excuse me using what has become something of a cliché; it had been a journey of discovery. The idea was to promote the book and then put my proverbial author's feet up for a while. But it wasn't too long before something started gnawing away at me. It wasn't that the book was incomplete or anything like that, but information had begun flooding into my mind on a regular basis relative to something that most of us take for granted - *light*.

The first thought to cross my mind was that I should publish a revised edition of *Transcognitive Spirituality* incorporating an additional chapter that would include this new information, but I knew that this would prove somewhat difficult because I wouldn't be able to simply add another section at the end of the book, it would have to be slotted in between existing chapters and that would interfere with the 'flow' of the work. I re-read the existing manuscript as it stood and it worked well, so to change it wasn't really an option.

I'm glad then, that I decided to write a completely new book especially in 'light' of the information that has come to me in one way or another since I began work in early October 2015. I had no idea then exactly how the narrative would develop, what information I would need to include or even the precise direction in which the book would go. I Just knew with some certainty that I had to get my ideas down on paper and what you are now holding in your hands is the finished article that I'm sure you will agree, deserves to stand alone in it's own right. That said, I do consider this an adjunct to the earlier book and if you haven't yet read that, I strongly suggest you do because I believe that it will make this one even more accessible to you.

I've structured *Light: The Divine Intelligence* into two sections - the Outer Light and the Inner Light because this book too, is a journey of sorts, one that will take you from what is commonly considered to be our external reality 'out there' to our internal reality 'in here'. Light in all of its forms (and it has many) is common to both.

As with *Transcognitive Spirituality* this information will stretch your mind somewhat, perhaps more for some than others and I would implore you to persevere with it, particularly through the earlier chapters, which, although written for the layman, are possibly more 'scientific' than you may have anticipated. This is purely because of the nature of the subject - light being quite a complex energy in so many ways.

What you will discover as you read the book is a movement towards the spiritual dimensions through the emergence of what I call the *inner light*. I have long maintained that science and spirituality are interconnected and the association between the official version of light described by science and the one referred to by mystics and seers becomes more obvious as each chapter unfolds.

As my writing continued across the weeks and months that I typed away, being led to discover the information that I needed to weave into the story whilst also being guided by my own intuition, it became even clearer to me, as I hope it will to you that this is a homage to the transforming power of light and the ultimate state of the evolved human condition - that of *enlightenment*.

Light is so much more than the energy that illuminates our world. It is informative, transformational and dynamic beyond measure. It connects deeply with everything on earth, not only providing the means for life to exist but also communicating with us in profound ways. If anything could truly be described as a 'messenger of the gods', then light would be it.

Writing this book proved to be a cathartic experience for me as well as a labour of love. I am my own severest critic and as I read through some of the passages that I had written I wondered if they were worthy of inclusion, totally relative to the story, or whether I had made them fit in a way that someone might force a jigsaw piece to go where it wasn't intended. But upon reflection and with much pulling of hair and wringing of hands I decided to leave everything in, even the parts that some of the more skeptical critics might decide stretch the boundaries of acceptability. I believe in presenting information that is worthy of consideration even if not all of the facts are verifiable. The reader will see that there are common threads uniting some aspects, most notably those concerning the claims of 'manifestations' made by followers of certain mystics, gurus, healers and prophets that are difficult to either prove or disprove.

However, at times we have to 'take a person's word for it' rather than expend energy searching for scientifically proven 'evidence' or verifiable facts. When we are dealing with the intangible and the subjective, empiric proof becomes harder to obtain. If a person says they have a 'gut feeling' about something, there is no way of disputing what they say - you have to accept that in all probability they are being truthful. It may be only later when certain events have transpired that what they said can be reassessed. Based on this fact and in the knowledge that I have done my research diligently I hope that you will read on with an open mind and enjoy the delights that light has in store for you.

Finally, as I hint in Chapter Seven, I could have made this a 'How to….' instructional type of book, so popular these days, especially with those who consider themselves to be upon the spiritual pathway. But I haven't for two very good reasons; firstly, as good as some of these books are, when it comes to spiritual growth there is no concise formula or foolproof set of instructions that can guarantee your life will be transformed. You might be able to learn the basics of strumming a guitar (as I have) or building a garden wall, but the foundation of true unfoldment takes a little longer to cement together. Secondly, enlightenment is not a fixed place and attaining it cannot be taught. It has to emerge from within, every single time. All that I or anyone can do is to reveal the signposts that point the way forward.

So what I have delivered is, I hope, a book that will not only give some direction and inspiration but most importantly, also shed a little light upon arguably *the* most incredible and wondrous phenomenon ever to emerge from creation.

Part One
The Outer Light

Light: The Divine Intelligence

1

...and the universe said 'Let there be light'

"There is only black light between the stars. It may seem that it's darkness, but it's really black light. There is no such thing as darkness. Darkness is a human concept. There's only black light between the stars."

Rama (1950-1998)

Our universe; magnificent, dynamic and extraordinarily beautiful and diverse has always been a source of fascination for me. As a small child I would often look up at the heavens, imagining faces in the clouds, convinced that they were actual people watching over me and after dusk my eyes inevitably scanned the myriad points of light scattered across the Milky Way, hoping to catch a rare glimpse of a shooting star or something equally spectacular. Even now I glance upward on clear, dark nights expecting to see the unexpected.

On cloudless days I still look skyward but because I am fair skinned and blue-eyed this results in me having to squint to protect my eyes from the bright glare of daylight and my father's warning to 'never look directly at the Sun Rob' has remained with me to this day. One of my earliest recollections of this comes from a time when, as an eight-year old he and I sat outside counting down the minutes to a solar eclipse armed with just a homemade pinhole projector fashioned out of two pieces of white card with a small hole in the centre. Through this invention we viewed the astronomical event without ever needing to look directly at the Sun and risk being blinded for life.

The Sun's light of course, as well as giving small boys an opportunity to experiment and interact with nature - what young lad has not risked starting a major fire by using a magnifying glass to focus the Sun's rays on a piece of paper or an empty crisp packet until it bursts into flames? - is necessary for all life on Earth. Ancient civilizations thought of light as being sacred and in many cultures the Sun was considered to be a god. Prior to Apollo, the Greeks referred to this as Helios whilst the Egyptian god, often portrayed as travelling through the waters of heaven in a boat, was known as Ra. When he flew across the sky it was day and when he moved through the underworld it was night. The Hindu god Surya was depicted riding across the sky in a horse-drawn chariot as was the Norse god Sol (Sunna). Countless other Sun gods or goddesses have also appeared throughout history to take prominence within cultures across the globe and it could be argued that even the Christian 'Son of God' - Jesus, began life as the Sun of God. In the biblical account our Sun appears in the third verse of the Book of Genesis which states that God, having created the heaven and the Earth spoke the words 'Let there be light', and there was light.

But what is light and how did the first light appear from darkness?

What is Light?

Light is part of the electromagnetic spectrum that, according to conventional science, ranges from radio waves (low frequency with a theoretical limit of 0 Hz) to gamma rays - the highest frequency so far detected, being in the region of 10^{27} Hz. These gamma rays come from galaxies and nebula other than our own but it is believed that cosmic rays may exist as high as 10^{30} Hz. Beyond these, scientists simply don't know although they do acknowledge that there is no known theoretical limit at either end of the spectrum. I would suggest however that these currently known extremes are only a very small fraction of the greater spectrum that is yet to be discovered.

Electromagnetic radiation waves are fluctuations of electric and magnetic fields that can transport energy between locations. What we refer to as visible light, although not inherently different from other areas of the electromagnetic spectrum is an extremely narrow bandwidth whose waves can be detected by the human eye. This is not by accident, but more by design because the detection of light provides living creatures with the means to explore their surroundings. A typical human eye will respond to wavelengths from about 390-700nm. In frequency terms, this corresponds to a thin band in the vicinity of 430-770 THz. Visible wavelengths pass through the 'optical window' a region of the electromagnetic spectrum that allows wavelengths to pass largely unchanged through the Earth's atmosphere and this window is also referred to as the 'visible window' because it overlaps the human visible response spectrum. Other known 'windows' such as the infrared lie beyond human vision and as such cannot be seen, although some creatures may be able to detect them.

Light has the characteristics of both particles (photons) and waves and the distance measurable between wave peaks is known as the 'wavelength'. Although we refer to contrasting ranges of the electromagnetic spectrum differently according to wavelength, everything can be considered to be part of one continuum of electromagnetic radiation *(Fig 1)*.

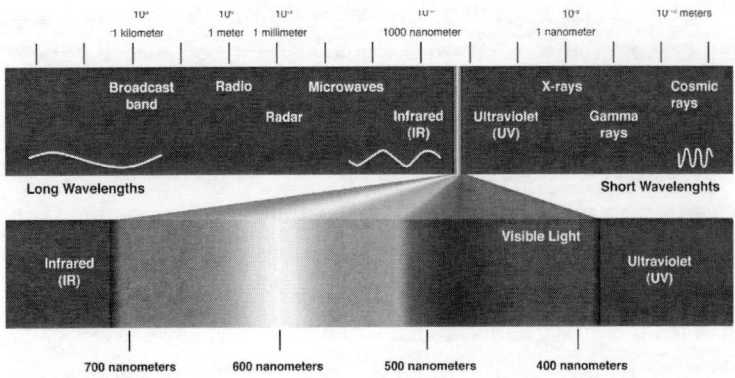

What we term 'sight' occurs when light enters through our eyes and undergoes a process known as optical refraction or bending, as it passes through the transparent structures of the eyeball. It is then transmitted to photoreceptor cells within the retina and this produces a chemical reaction, creating an electrical stimulus that travels along the optic nerve and the visual pathways to the primary visual cortex in the brain.

The light rays form an image on the retina and the amount of electrical impulses transmitted to the brain are interpreted allowing us to 'see'. So, we actually 'see' with our brain and various parts of the brain, memory for example, are then involved in the process of perception. A modern alternative that describes our ability to interpret light, or indeed anything illuminated by light is the expression to *decode* and this is a term that I will be using extensively throughout this book for reasons that will become clear.

When we decode something we are essentially translating or unscrambling a signal or message (information) from one state to another. If our human brain was not sufficiently developed to enable the decoding process to occur, we would not be able to see anything at all and would effectively be blind. So because our brains are able to decode the area of the electromagnetic spectrum that covers visible light, this is the 'reality' that as humans, we experience.

In pure scientific parlance electromagnetic radiation can also be described in terms of a stream of photons - elementary particles exhibiting wave/particle duality moving at the speed of light *(Fig 2)*. A photon is the smallest known quantity of energy (quantum) that can be transported. It was the discovery that light travelled in discrete quanta that lead to the evolution of 'Quantum Theory'.

Physicists have speculated that photons have no mass, although they do have some characteristics of other particles but their frequency is independent of the influence of mass. Also, they don't carry a charge.

Light: The Divine Intelligence

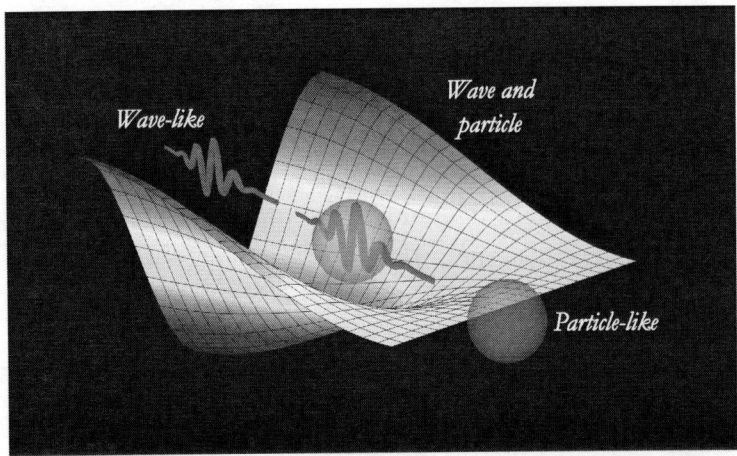

Fig 2. From the waveform to the particle

Photons influenced Albert Einstein's theory of relativity allowing us to understand the importance of the speed of light and with it the knowledge of the interaction of time and space *(spacetime)* that it produced.

It was though, the Danish astronomer Ole Roemer (1644-1710) who, in 1676, first came close to accurately measuring the speed of light. His method was based on observations of the eclipses of the moons of Jupiter, particularly Io, the innermost moon (closest to Jupiter). Whilst observing Io, Roemer noticed that the times of its eclipses seemed to depend on the relative positions of Jupiter and Earth. If Earth was close to Jupiter, the orbit of Io appeared to speed up. If Earth was far from Jupiter, it seemed to slow down. Reasoning that Io's orbital velocity should not be affected in this way he deduced that the apparent change must be due to the extra time for light to travel when Earth was more distant from Jupiter. Roemer thus estimated the speed of light to be 140,000 miles per second, which considering the methods he employed was remarkably good.

Across the centuries this measurement has become more accurate and it is now known that the true speed of light in a vacuum is 186,282 miles per second (299,792,458 metres per second). Possessing this knowledge is vital in helping scientists to determine the movement of stars, planets and entire galaxies as well as to speculate on the very nature and age of the universe itself.

An interesting phenomenon occurs when a source of light moves. If a light source, such as a star or galaxy moves away from an observer the wavelength extends and in the case of the visible spectrum, the light shifts towards the red portion of the spectrum - hence the term *red-shift*. Astronomers often use the term red-shift when describing how far away a distant object is. To understand what a red-shift is, think of how the sound of a siren changes as it moves toward and then away from you. As the sound waves from the siren move toward you, they are compressed into higher frequency sound waves. As the siren moves away from you, they are stretched into lower frequencies. This shifting of frequencies is called the *Doppler Effect*. A similar thing happens to light waves in the visible portion of the electromagnetic spectrum. When an object in space moves toward us its light waves are compressed into higher frequencies or shorter wavelengths and we say that the light is blue-shifted.

When an object moves away from us, its light waves are stretched into lower frequencies or longer wavelengths, and we say that its light is *red-shifted (Fig 3)*.

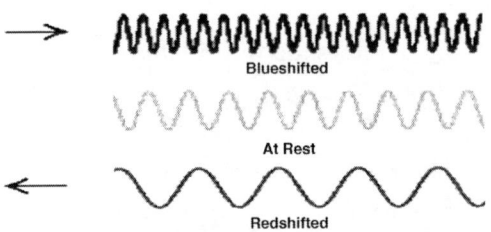

Fig 3 Doppler waves

It was a scientist named Edwin Hubble (1889-1953) who is credited with being the first astronomer to notice the galactic red-shift. In 1929 during his extensive observations Hubble realised that almost all of the galaxies he studied exhibited red-shift properties, suggesting that the universe was expanding. Furthermore, the more distant the galaxy, the faster they seemed to be moving. This discovery was of great importance to the astronomical world, overturning the conventional view of a static universe and came to be known as Hubble's Law. To understand the principle of an expanding universe more fully a useful analogy would be to take an empty balloon, draw dots all over it to represent galaxies, and pretend that we live on one of the dots. As you blew up the balloon you would witness the entire number of dots move apart from each other and the ones that are farthest away from us move the fastest. The difference between this analogy and the actual universe is that although the galaxies are being pulled away from each other by the universe's expansion, they are not 'stuck down'. If we were to replace the dots on the balloon by groups of ants this would give a better analogy for this idea. Astronomers refer to the velocity that a stuck down galaxy would have as its 'Hubble Recessional Velocity' and any deviation from this speed as its 'Peculiar Velocity.'

There are exceptions, but generally almost all of the galaxies are red shifted; they are moving away from us, due to the Hubble expansion of the universe and there are only handful of nearby galaxies that are blue-shifted. In addition to the apparent motion resulting from universal expansion, individual galaxies also have their own intrinsic or peculiar motions; i.e. each galaxy is in motion irrespective of the universe's expansion and has its own unique velocity. These velocities are in the order of hundreds of kilometres per second and in regions close enough to our own galaxy where the Hubble expansion results in less outward expansion than this, the galaxies' peculiar velocities (if they are large enough and sufficiently towards us) can overcome that expansion, resulting in a blue-shift.

Relative to our quest to discover the origin of the first light to appear in the universe, Hubble's Law is crucial and the understanding that the universe is expanding, even to this day, suggests that it was once very small, perhaps even originating from a single particle.

The hot universe

All objects emit electromagnetic radiation continuously and

the wavelength of radiation or light depends on the temperature of the object. If the universe was once condensed into a single particle then the intense compression of all substances and energy would generate enormous heat and corresponding radiation. As the universe expanded following the so-called *Big Bang* or as I would suggest an equivalent event such as the ejection of extremely hot, condensed matter from the singularity or *Unified Field,* the apparent temperature of this light would have cooled down. As it happens, this is precisely what scientists discovered. Astronomers Arno Penzias and Robert Wilson were using a large horn antenna in 1964 and 1965 to map signals from the Milky Way, when they unexpectedly discovered what has come to be known as *Cosmic Microwave Background Radiation* (CMB) or 'noise' to you and me. Working for Bell Labs in New Jersey, USA Penzias and Wilson were given the opportunity to use its horned shaped antenna with an obsolete satellite system called Echo to analyse radio signals from the spaces between galaxies. But when they started using it, they encountered a persistent noise of microwaves coming from every direction. They soon realised that if they were to continue to conduct experiments with the antenna, they would have to find a way to delete the static. Having done everything they could to remove the interference, including clearing out the droppings from pigeons that were nesting inside the antenna (and the pigeons too) the noise remained. At the same time, the two astronomers discovered that Princeton University physicist Robert Dicke was suggesting that if the Big Bang had indeed occurred, there would be low-level radiation to be found throughout the universe. Dicke was in the process of designing an experiment to test his hypothesis when he was contacted by Penzias and quickly realised that the two men from New Jersey had beaten him and his colleagues to the draw. Although both groups subsequently published their results in Astrophysical Journal Letters, it was Penzias and Wilson who received the Nobel Prize for the discovery of Cosmic Microwave Background Radiation.

Found everywhere throughout the cosmos the CMB radiation had cooled to an apparent temperature of 3K (Kelvin, the international unit of heat) or 270°C and this is, according to mainstream science, the oldest known light in the universe.

But is it?

Dark Energy and Dark Matter

As I've already explained, light waves, also called radiation, carry energy. If you step outside on a hot, sunny day you will immediately feel that energy interacting with your skin. Einstein's famous equation, $E = mc2$, states that matter and energy are interchangeable, being merely different expressions of the same thing and of course if we consider again our nearest star, the Sun, it too is powered by the conversion of mass to energy. But energy is supposed to have a source - either matter or radiation. Yet space, even when devoid of all matter and radiation, appears to have a residual energy. This 'energy of space,' when considered on a universal scale, leads to a force that increases the expansion of the universe. So what is it and where does it originate?

Firstly we need to distinguish between what is referred to as *Dark Energy* and *Dark Matter*. Scientists regard these as being hypothetical, in that they have not yet been proven to exist according standard scientific protocol. Neither has it been proven that they are manifestations of the same thing, but *Dark Matter must exist to account for the gravity that holds galaxies together*. It seems that if the only matter in the universe was that which scientists could directly detect, galaxies would not have had sufficient matter to have formed in the first place. The galaxies that are observed today would also fly apart because they would not have enough matter to create a strong enough gravitational force to remain intact.

Dark Energy must also exist to account for the rate of expansion that is observed throughout the universe. As we've already discussed, not only is the universe expanding but it is also accelerating. Gravity's force, although stronger when things are close together, is weaker when they are far apart and because gravity is weakening with the expansion of space, Dark Energy now composes over two thirds of all the energy in the universe (Fig 4).

I propose that not only are Dark Energy and Dark Matter essential components of what we consider our 'physical universe' but that they are also aspects of the same source from which the first light originated along with all other frequencies or dimensions of 'reality' - upon which I will elaborate further in part two of this book. Why can't we 'see' this energy or matter? - because it is outside of the visible portion of the electromagnetic spectrum and most certainly beyond the parameters of the normal decoding mechanism of the human brain. So often we are only aware of something because of the *effect* produced by its presence. An obvious example would be the effect of the wind moving the leaves and branches of a tree; we cannot 'see' the movement of the air because it is outside our decoding range, but we do 'see' and very often 'hear' the phenomena it produces. In the same way Dark Energy and Dark Matter are detectable only because of their effect on the visible matter around them and together they make up most of the universe. According to consensus among cosmologists, Dark Matter is composed primarily of a not yet characterised type of subatomic particle. The search for this particle, through a variety of means, is one of the major efforts underway in particle physics today.

Perhaps though, the term 'dark' is a misleading one. Just because something is labelled in a certain way doesn't make it thus. Change the label and you change to a degree the perception of it. We humans have a tendency, through our conditioning, to think of something dark as being a bit dodgy, perhaps mysterious or possibly evil in some way. But change the name to 'transparent' or 'etheric' and the emphasis becomes much more accessible and even, dare I say it, metaphysical.

Light, the giver of life

Light is also fundamentally linked to our perception of time - not just in the broadest sense, relating to stars, galaxies and the age of the cosmos, but in very personal and intimate ways that affect considerably our ability to function on this planet. In addition it has great influence in determining our sense of time, or as some might speculate, the *illusion* of time, with far reaching implications. However, although we have already touched briefly upon the concept of both time and spacetime it is not my intention here to delve into the physics that underpin these because there are places far more relative and immediate to the story of light awaiting our exploration. Readers wishing to know more about these aspects will though, find them covered in greater depth in my book *Transcognitive Spirituality - Shaping the New Paradigm at the Interface of Consciousness and Reality* (to give it's full title) For now we will focus on the knowledge that everything upon our Earth and throughout the immediate vicinity of the solar system is influenced to one degree or another by light and in particular, our nearest star, the Sun.

Simply put, conventional science postulates that the Sun produces energy by the fusion of hydrogen into helium deep within its core. Like most stars, the Sun is composed mainly of hydrogen gas and the process of producing energy starts from the hottest area, the core and is so compressed that enormous amounts of hydrogen atoms fuse together. In the process known as nuclear fusion, the hydrogen atoms are converted into helium, the same gas used for balloons. Nuclear fusion produces massive volumes of energy that radiates outward to the surface of the Sun and beyond into the surrounding space, continuing to taper off and weaken as it makes its way to the Earth's atmosphere. It eventually reaches us in the form of sunlight, a combination of radiant light and heat.

It is worth mentioning here that some scientists have hypothesised the Sun as being electrically powered. In August 1972 Ralph Juergens (1924-1979) inspired by Immanuel Velikovsky's (1895-1979) contention that electromagnetic forces played a crucial role in sculpting the surfaces and shaping the orbits of the bodies of the solar system, published a paper in which he stated:

"....the interplanetary medium suggest not only that the Sun and the planets are electrically charged, but that the Sun itself is the focus of a cosmic electric discharge - the probable source of all its radiant energy..."

If we consider for a moment the emerging *Plasma Universe Paradigm,* and its many far-reaching implications, the idea of the 'electric Sun' becomes more plausible. Mainstream science, for the most part, looks on the universe as being electrically neutral and purely mechanical; a place where the weak force of gravity holds sway. *Plasma Cosmology* by contrast, acknowledges the electrodynamic nature of the universe and recognises that gravity and inertia are not the only forces at work.

Plasma, (like hypothetical Dark Matter) exists abundantly throughout the cosmos and at least ninety-nine percent of the *known universe* is, in fact, matter in its plasma state. Some scientists even go as far as to suggest that it was the first state of matter. Plasma in space consists entirely of ions and electrons, and is thus very energetic or 'hot' *(Fig 5)*. Only when cooled does it form the matter to which we are familiar here on Earth - solids, liquids and gases. Because plasma remains electrically charged in space, it is influenced more by electromagnetic forces than gravity.

Solid	Liquid	Gas	Plasma
Example Ice H_2O	Example Water H_2O	Example Steam H_2O	Example Ionized Gas $H_2 \rightarrow H^+ + H^+ + 2e^-$
Cold T<0°C	Warm 0<T<100°C	Hot T>100°C	Hotter T>100,000°C (>10 electron Volts)
Molecules Fixed in Lattice	Molecules Free to Move	Molecules Free to Move, Large Spacing	Ions and Electrons Move Independently, Large Spacing

Fig 5. Plasma - the first state of matter?

The importance of the theory of an Electric Universe and the recognition of plasma energy is important in helping us to understand the way in which the Sun's energy reaches Earth and one of the most enduring mysteries in solar physics is why the Sun's outer atmosphere, or corona, is millions of degrees hotter than its surface. Now scientists believe they have discovered a major source of hot gas that replenishes the corona;

jets of plasma shooting up from just above the Sun's surface.

But here is another explanation; just as we decode information into electrical, digital and holographic states, so too does the universe exist at these different levels of expression. What if the Sun was not a massive nuclear reactor, generating light and heat from its core, but was in fact drawing its power from the electrical level of the universe? What if the Sun and all stars are electrical processors and transformers, tapping into the electrical field of the universe and turning its power into light in much the same way as a light bulb?

Nobel prize winning scientist Irving Langmuir (1881-1957) discovered that when two plasmas, each with a different electrical charge meet, an energetic barrier is created between them. These barriers are now known as 'Langmuir Sheaths' because they describe planetary energy fields, also known as magnetospheres. The Sun has its own version called the Heliosphere *(Fig 6)*.

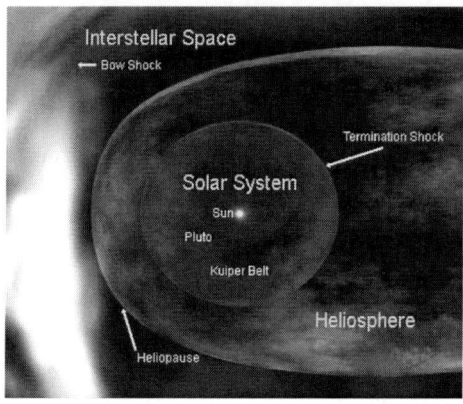

The Heliosphere is defined as the region of space surrounding the Sun, in which the solar wind, the solar magnetic field, and all of the ejections of matter from the Sun, play a large role in controlling how plasma inside the solar system behaves. Langmuir also discovered that plasma appears to have the ability to self-organise as a response to electrical change and physicists have since created blobs of gaseous plasma that can grow, replicate and communicate - fulfilling most of the traditional requirements for biological cells. This is no surprise really, because like every aspect of creation, *plasma has consciousness.*

Another scientist, Kristian Olaf Birkeland (1867-1917) who was the first to explain the phenomenon of the Aurora Borealis, also discovered that when electricity passes through plasma, filaments are created. In 1969 when field-align currents had been identified in the Earth's atmosphere, they were named in his honor - 'Birkeland currents.' They are caused by the movement of a plasma perpendicular to a magnetic field and often show a twisted "rope-like" magnetic structure. They are also known as magnetic ropes and magnetic cables. Interestingly, Birkeland currents are also one of a class of plasma phenomena known as a 'z–pinch' so named because the peculiar magnetic fields produced by the current pinches it into a filamentary cable. This can also twist, producing a 'helical pinch' that spirals like a twisted or braided rope. Those readers who are familiar with the spiral appearance of the DNA double helix will see the obvious similarity here. It is also worth noting, in relationship to light, that electrons moving along a Birkeland current may be accelerated by a plasma 'double layer' (a structure in a plasma which consists of two parallel layers with opposite electrical charge). If the resulting electrons approach relative velocities (the speed of light) they may subsequently produce a 'Bennett pinch' which in a magnetic field will spiral and emit electromagnetic radiation known as 'synchrotron radiation' that includes radio, optical (light), X-rays, and gamma rays.

Lightning

As we will see, not all light comes from our Sun, the stars or other cosmic phenomena. Light is everywhere and no more so than in the awesome lightning displays that we witness frequently across the globe. A recent paper published by researchers working at England's University of Reading cited evidence indicating that high-speed 'solar winds' *(Fig 7)* play a major role forming lightning storms in the Earth's atmosphere as well as determining their frequency and severity. Cosmic rays (supernova-generated energised particles from deep space) are already known to play a similar role, although to a lesser extent because, in order to reach the Earth's atmosphere, cosmic rays must first penetrate the solar wind, which is a continuous outgoing stream of plasma (ionised electrons and protons) released from the Sun's upper atmosphere. Flowing outward at supersonic speeds to enormous distances, the solar wind creates the gigantic semi-porous 'bubble' (Heliosphere) that helps protect our Earth and solar system from the harsher elements of the surrounding interstellar medium. Because the solar wind's one-million-mile-per-hour particle stream varies in density, temperature and velocity, its energising effect on the Earth's ionosphere (part of the Earth's outer atmosphere) also fluctuates and the more electrified it becomes, the greater the likelihood that influxes of charged particles will filter downward and become engaged in the complex process of lightning storm formation.

Fig 7.

There are several types of lightning *(Fig 8)*. including 'cloud to ground' (CG), 'cloud to air' (CA), sheet lightning or 'intra cloud lightning' (IC), 'cloud to cloud'(CC) and different variations according to prevailing conditions that include those that originate high in the atmosphere and are seldom witnessed visually. These are referred to as 'trans luminous events' (TLEs) and include 'red sprites', 'blue jets' and 'elves'.

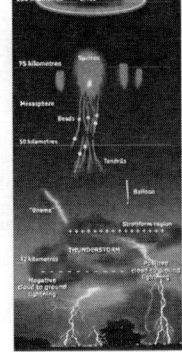

Fig 8.

Scientists discovered "a substantial and significant increase in lightning rates" for up to forty days after solar winds hit Earth's atmosphere. The reasons behind this are not yet fully understood, but it is thought that this could be because the air's electrical charge alters as the particles (which are themselves electrically charged) hit the atmosphere. If proven, this could give weather forecasters another string to their bow, allowing them to incorporate into their predictions, information about solar wind streams that are being watched from space. This stream of particles would change with the Sun's twenty-seven-day rotation, and researchers hope this could improve long-range forecasts.

The Sun influences the climates of all the planets in the solar system, but significantly our own Earth. It is the source of most of the energy that drives the biological and physical processes in the world around us. It fuels plant growth in the oceans and on land that form the base of the food chain, and in the atmosphere it warms the air that drives our weather. Over longer time-scales changes in solar intensity are a critical factor in influencing climate. One only has to look back through history to periods when the Earth was much colder or much warmer to see how plant and animal life were affected. Even small changes in solar activity can produce dramatic changes. But why is light in particular so significant to life?

If like me, you enjoy a spot of gardening, you will be acutely aware that unless you were born with green fingers, plants are very vulnerable to fluctuations in temperature, moisture and light, all of which are linked. Plants get energy from light through a process called *photosynthesis*. This is how light affects the growth of a plant.

Without light, it would not be able to produce the energy it needs to grow. Not all plants are the same of course. Some prefer more shade and others need light that falls more into the 'blue' spectrum of the light scale. Commercial growers sometimes use artificial 'grow lights' or fluorescent lights to achieve this, whilst recognising that incandescent and halogen lights are more of the 'red' spectrum and so, less conducive to good growth. Also, the length of time that plants are exposed to light, particularly full spectrum sunlight, impacts their wellbeing. The phenomenon of photosynthesis cannot be over emphasised in terms of its importance to life on Earth. As many school children will know, it is a chemical process through which plants, some bacteria and algae, produce glucose and oxygen from carbon dioxide and water, using only light as a source of energy. Through this process glucose is used to release energy that the plant needs for other life processes. The plant cells also convert some of the glucose into starch for storage and this can then be used when the plant requires it. This is why 'dead' plants are used as biomass, because they have stored chemical energy in them. This is also why the recommended 'five-a-day' is good for us because when we eat fruit or vegetables, our bodies extract this energy and use it to keep us alive and healthy.

Glucose in plants is also needed to make other chemicals such as proteins, fats and plant sugars that are all necessary for it to carry out essential growth and other life processes. Most leafy plants have evolved to have leaves that are broad, flat and exposed to capture as much of the Sun's energy (sunlight) needed for photosynthesis. The process takes place inside plant cells in small structures called *chloroplasts*. Chloroplasts contain a green substance called chlorophyll and it is this that absorbs the light energy needed to make photosynthesis happen. It is important to note that not all the colour wavelengths of light are absorbed, particularly from the green spectrum. Plant leaves usually have a large surface area, so that it can absorb a lot of light, its top surface being protected from water loss, disease and weather damage by a thin waxy layer. The upper part of the leaf is where the light falls, and it contains a type of cell called a *'palisade cell.'* This is adapted to absorb large amounts of light and has lots of chloroplasts. The more light there is, the more photosynthesis that occurs (although too much light may be detrimental) enabling, through further interactions with water and air, growth to occur. There are some plants that thrive with little or no light, but these are few and far between.

In understanding the importance of light in the life process of plants we can more readily see the obvious connotations with all other living forms upon Earth, including our own. We can recognise how this energy, to which everything is connected and upon which everything depends is without doubt, the giver of life. Without light, there would be only darkness and in darkness, very little can survive. Even creatures living in the deepest oceans, where no direct sunlight can penetrate, must find ways of surviving, either by creating their own internal light (bioluminescence) or by utilising internal chemical processes as the basis for food. The process behind this is nothing short of magical. Even some land based creatures have the ability to create light, the most familiar being fireflies, flashing to attract mates on a warm summer night, but there are others too, including glow-worms, some millipedes, a particular species of snail, and even a certain type of fungi known as foxfire.

It is in the oceans though, that we witness the most astonishing light show with an assortment of creatures displaying their bioluminescent ability. The list is almost endless and includes both fish and invertebrates. Indeed, of all the known groups of organisms possessing the means to generate light, over four-fifths live in the ocean. You may ask the question 'why would any species develop the ability to generate light, particularly those considered as food for predators?' Well, here are a couple of reasons; firstly, a sudden burst of light may startle a predator, giving the prey a chance to escape. A deep-sea squid, for example, can emit a flash of light before speeding off into the gloom, whilst a species of worm known as 'green bombers' can throw their 'light grenades', before disappearing into the darkness, leaving the predator totally distracted by the light. Secondly, the chief defence for some creatures is not to run away, but to negotiate! Based on the ancient principle, 'the enemy of my enemy is my friend', emitting light may be one way for the prey to call the in cavalry or invite the 'predator of the predator'. This may be especially useful for species whose swimming capability is not too good.

Some sea creatures, especially those existing in the open expanses of the ocean have also developed the ability to be transparent with the obvious advantage of being more difficult to spot. There is a snag though, creatures that are 'see through' and who eat something that is glowing may suddenly become visible! Perhaps that is why so many of them have developed guts that are opaque.

So if light can be utilised so effectively by plants, land-based and sea creatures, what about us humans? Well, we've managed to harness the power of the Sun quite well it would seem, drawing on solar power in numerous ways, as we have too with our generation of electricity, without which the modern world would cease to function. In fact, at times we create so much light that we even experience negative effects such as light pollution and if you have ever flown in an aircraft at night and glanced through the window when flying over a large city, you will understand exactly what I mean. From high above the Earth, it is easy to distinguish the human footprint, standing out from the surrounding countryside, simply by the light emitted from street lamps, illuminated signs and motor vehicles and down here on terra firma, the artificial light from our towns and cities sadly prevents us from seeing many of the stars within our own galaxy. But we would be all the poorer without it. Sitting beneath the night sky around the light and warmth of an open fire may be a romantic thing to do, reflecting upon our own ancestry and transporting us back to distant times when this was the only light we had. But then, reality dawns and we long for a hot shower, a warm bed and a good book, read by courtesy of the light from our bedside lamp. Oh, and should we need to catch up with the evening news – we'd better turn on the TV.

Despite the creature comforts that our faithful servants, the energy of electricity and its attendant visible light bring us, there is much more going on with these two forces, both around us and within us, that we are not consciously aware of. Our 'physical' body (the form that our consciousness decodes into what we perceive as physical form) is 'hot wired' to transform electromagnetic energy and most importantly, *the information contained within it* into various other forms of energy that we can utilise to beneficial effect. As we will discover, we are both transmitters and receivers of light, at every level of our being.

2

Ghost in the Machine

> *"Light is important to us humans. It influences our moods, our perceptions, our energy levels. A face glimpsed among trees, dappled by the shadows and the green-tinged light reflected from the forest, will seem quite different to the same face seen on a beach in hard, dry, sunlight, or in a darkening room at twilight, with the shadows of a venetian blind striped across it like a convict's uniform."*
>
> John Marsden (1950 -)

Let us for a moment consider the human body somewhat differently than we are probably used to. Conventional science tells us that after fertilisation, 'life' begins and our physical body slowly emerges in our mother's womb over a period of approximately forty weeks, growing and developing continually. In this time the fetus metamorphoses from an embryo of multiplying cells into a being with organs and muscles that is ready for life outside the womb. The pregnancy is divided into three developmental phases or trimesters. The first trimester lasts from the first through the thirteenth week and is vital for the fetus because by the end of the third month it will have developed all of its organs. Around the fourteenth week the baby's arms have almost reached the final lengths they'll be at birth and the neck has become more defined. The spleen has also seen the development of red blood cells. The baby's sex will soon become apparent and for females, ovarian follicles begin forming whilst for males, the prostate appears and by the fifteenth week, the skeleton has begun forming. Growth continues through to the third trimester, which begins around the twenty-eighth week and lasts until birth. During this phase development continues at a pace, with many vital aspects emerging, including the digestive system and crucially the brain, until finally, if all is well, the child is born. This 'miracle' occurs at a phenomenal rate across the world, with latest estimates suggesting that 350,000 babies are born every day.

Conception and birth are a part of life that some take for granted, but this, we should never do and despite its wide-ranging capabilities science has yet to understand what life is, when it begins or from whence it originates. However, current scientific thinking regarding when 'life' is considered to commence, falls into five categories, which are:

- **Metabolism** - the commencement of cell activity
- **Genetic** - a genetically unique person begins at conception with the fertilised egg containing the complete genome
- **Embryology** - the beginning of gastrulation, about fourteen days after fertilisation
- **Neurology** - the onset of EEG activity
- **Ecology** - when the foetus can sustain itself outside the mother's womb

This doesn't necessarily mean that there are five possible 'points' to choose from. The reality is much more complex and these are not so much five different points, but rather five distinct criteria leading to five different areas of change that *could* be defined as 'the start of life'. One of the main viewpoints and the one that I favour is that there is no one point where life begins. Instead, the beginning of life is a continuous process. Furthermore, there is one other distinct category that I believe should be added to the list:

- **Metaphysical** - defined by non-physical intelligence (soul) connecting with the fertilised cell at the moment of conception

This sixth point may be uncomfortable for the majority of mainstream scientists but requires only a small shift in thinking when considering what we currently know about the quantum world. So let's turn this on its head and look at things from an alternative perspective:

- Life already exists beyond the range of our physical senses
- Life is not confined to the 'physical' universe but operates through it
- Physical 'reality' is a persistent illusion created by the mind
- Matter is not solid but appears that way because of the resistance between two electromagnetic states

- Our mind/body 'computer' decodes information from the waveform states, thus allowing our real nature (infinite consciousness) to experience 'the world' through a 'physical body'
- Our unique individual consciousness (life) decodes vibrational information from the metaphysical universe into other forms of information; Electromagnetic, Digital, Holographic
- Decoding 'reality' can be compared to logging on to the Internet. We observe and interact with it through a computer. Our *bodymind* acts as a computer for our consciousness and provides for us the means to 'log on' to the waveform information construct. By doing this the five main human senses decode vibrational information into electrical information which is then transmitted to the brain and our entire genetic structure where it is decoded into digital and holographic information that we experience as 'physical reality'

(see 'The Map' illustration at the front of this book)
- The human body acts like a lens that focuses our attention upon a specific small range of frequencies
- We are *individualised consciousness*, an aspect of *infinite consciousness*, operating through mind
- Through the holographic nature of creation every part reflects the whole

Sort of puts a different slant on life doesn't it? What I'm suggesting here is the antithesis of current mainstream thinking; that life is not some kind of cosmological accident or dependent on anything pertaining to the physical universe. Rather, it is the creator of the latter, illusory reality, functioning both within and through it at every level. When viewed in this way we can begin to formulate a different set of criteria about the nature of 'self' as it appears throughout each unique human form and the mechanism employed to facilitate this.

From the perspective of a 'soul' about to incarnate into a newly fertilised egg we can speculate upon the following sequence occurring:

- Depending upon the level of spiritual development, the incoming individual, operating through its own freewill, chooses the vessel into which it is about to incarnate. This may also involve an earlier assessment of the lineage as well as the purpose of the future life
- The light of individualised consciousness (an aspect of the soul) connects with the fertilised egg (either within the womb or within a laboratory test tube
- A non-physical holographic 'matrix' is formed at a higher vibrational waveform information level, unique to the individual. This can be considered as a kind of 'blueprint' or mould for the material form
- The light of consciousness of the incarnating individual begins operating through the waveform matrix, decoding information from its own innate higher self in combination with that already existing within the genetic formulation of the egg, the latter placing certain restrictions upon the overall formulation

- The combination of higher (faster frequency) and lower (slower frequency) energies allows the individual mind to decode the waveform information into 'physical form' - a process that continues as the embryo develops
- Around and during the three trimesters, formulation of the 'physical' body continues and quickens as the brain develops, allowing the mind to decode from the waveform information construct ever-greater levels of vibrational information into matter
- The process completes to the point where 'birth' occurs and the individual consciousness, although still greater than the incarnated form, functions through the lens of the mind and the limitations imposed upon it by the hereditary, genetic structure through which it experiences

Insofar as light in concerned, although the womb is dark, the developing fetus is sensitive to light. At around the eighth week of pregnancy the eyes and the retina - the layer of cells at the back of the eye that perceives and processes light - have begun to form. By week sixteen those cells have begun to actually pick up on light. Even though the eyelids can't yet open, the eyes can make slight movements from side to side in response to light. By week twenty-six, or the end of the second trimester a baby's eyes are just about as fully formed as they are going to be. Not only can they sense light but they can finally open and even blink.

Interestingly scientists have also discovered something else that helps an unborn baby's eyes - sunlight. Some photons of light still make it through the mother's skin, especially if she is outside in the sun and just that small amount of light, it turns out, is key to helping the eyes develop. Experiments have shown that when pregnant mice are kept in total darkness, their offspring are more likely to have vision problems after birth. Similarly, research has also revealed that babies of women who live at northern latitudes and become pregnant during the darkest months of the year are at an increased risk of certain eye disorders.

Clearly, there is much more to light than we have yet discovered. But further clues can be found deep within the human body itself, connecting us to each other and to light sources existing within and beyond the known universe. One of the most important, perhaps *the* most important, lies deep within the brain and is known as the *pineal gland (Fig 9)*. Little appears to be known about this incredible pine-cone shaped organ, once dubbed 'the third eye', although the Egyptian civilisation seemed to be somewhat aware of its potential *(Fig 10)*. But on the physical level at least, it is responsible for the production of the seratonin derived hormone, *melatonin*. This is significant because its secretion is dictated by light and researchers have determined that melatonin has two primary functions in humans - to help control our *circadian rhythm* and also to regulate certain reproductive hormones. If you are unfamiliar with it, the circadian rhythm is a twenty-four hour biological cycle essentially characterised by sleeping and waking patterns. The pineal gland is connected to the rest of the hormonal system and the production of melatonin also influences greatly the functioning of other parts of the body. For example, when it's dark and we are asleep, melatonin modifies the secretion of hormones from organs such as the *pituitary* - considered as the master gland of the hormonal system. This in turn regulates the secretion of hormones controlling growth, milk production and egg and sperm production. Crucially it also regulates the action of the *thyroid gland*, which is concerned with metabolism, and the *adrenal glands,* which control the excretion of bodily waste. It is obvious then that fluctuations in light and darkness according to the seasons of the year will influence rhythms of growth, reproduction and activity in both humans and animals. Just as animals and plants perceive the shortening of the days in the autumn and sense the onset of winter (and the reverse in spring and summer) so do we humans if we are honest about it. The trouble is though, that because we are so indoctrinated to working for and obeying 'the system' and so conditioned to social norms, (in contrast to our instincts) we get up before sunrise and often do not retire to bed until way after sunset, which isn't ideal for us. Animals though, follow their instinct, as can be seen more readily in those that hibernate during the colder, shorter days of winter and are all the better for it.

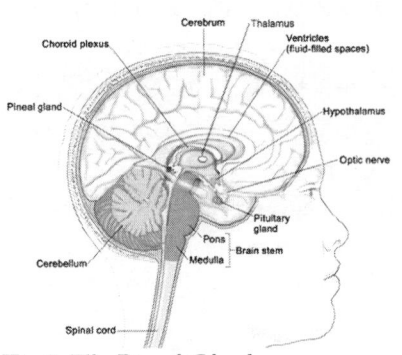

Fig 9. The Pineal Gland

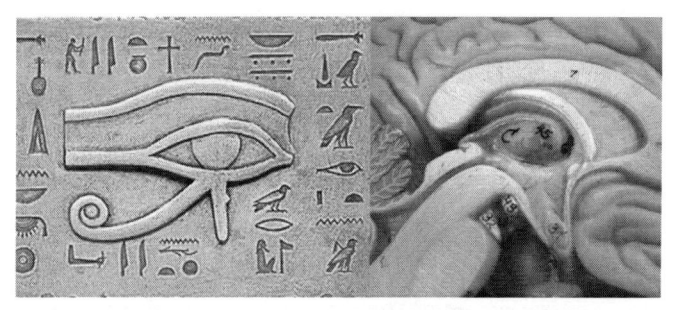

Fig 10. The ancient Egyptians were experimenting with altered states of consciousness by using the pineal gland to explore 'the realm of the spirits'.

The Pineal interface

The question arises; why would a race so advanced as the Egyptians place so much emphasis on the pineal gland and what did they know about its functioning that modern science, seemingly does not? References appear throughout Egyptian archaeology and in other cultures too. It was also known as the *Eye of Horus*. This symbol was passed down through the ancient mystery teachings and can still be found on the American dollar bill, located at the top of the pyramid structure on the back of the dollar note, although in this context it may carry a different interpretation.

The Eye Of Horus is an Egyptian symbol of protection, royal power and good health. Horus was the ancient Egyptian sky god who was usually depicted as a falcon, and his right eye was associated with the *Sun God RA (Fig 11)* who ruled in all parts of the created world: the Sky, the Earth and the Underworld.

The Egyptians worshipped the Sun and it was seen as the ruler of all of creation. RA is always shown with a sacred Sun disk on his head representing enlightenment and the connection to divine intelligence.

The pineal has also come to be known as the 'All Seeing Eye' (or as the third eye or the spiritual eye) and is often referred to as 'the seat of the soul'. It supposedly remains dormant until the human soul reaches a certain vibrational and spiritual level at which point the gland is activated by the light of the higher self, signalling the divine energies from the *Kundalini* (root chakra) to rise, thus activating the human chakra system.

Fig 11. RA, the Egyptian Sun God

Can it be that the pineal gland is more than simply an integral part of the endocrine system? Is it that this tiny organ is capable of far more than modern science is willing to admit to? I don't think there is much doubt of this. To me it makes sense that consciousness would evolve a number of arrangements and a clear *modus operandi* to enable the higher and lower, or finer and denser states of being to communicate. In modern parlance, this is referred to as an interface. Some might suggest that this has been adequately taken care of via the brain/mind combination, but as I will illustrate, there are many such interfaces throughout the physical body, most of which employ light in varying degrees.

The mind – body problem

The classic philosophical mind-body problem centres around the difficulty of explaining how mental states, events and processes - like beliefs, actions and thoughts - are directly related to physical states, events and processes, given that in conventional thinking, the human body is a physical entity and the mind is non-physical. Simply put, how does something non-physical, result in the movement of something physical? Many philosophers have grappled with this issue across the centuries with most solutions being either dualist of monist in nature. Dualism maintains a rigid distinction between the realms of mind and matter, whilst monism maintains that there is only one unifying reality, substance or essence in terms of which everything can be explained. I favour the latter, because my understanding of the *holographic* nature of life, in which consciousness underpins every facet of creation, suggests to me that this is so. It then makes perfect sense if, within the lower, denser aspect of being, a mechanism exists to act as a 'go-between', allowing the free flow of energy from one state to another. As electricity requires a conduit to reach the light switch, so too does light utilise human circuitry to move freely through the body and the pineal gland through its connectivity with the endocrine system offers one such means. As I wrote in my last book:

'Our connections to other frequencies whilst we are restricted to decoding reality at the 'physical' level are further enhanced through additional energetic mechanisms that are worthy of mention. These are the electromagnetic fields surrounding and interpenetrating us that are known collectively as the 'aura',

the system known as the 'chakras' and the powerful circuitry that acts as 'the motherboard' of the physical body, the 'meridians'. Each of these arrangements helps facilitate the flow of information between the vibrationary energy levels of which we are composed, ensuring a healthy balance is maintained.'

It is in our best interests to understand something of the nature of our own individual constitution and to recognise that we are composed of far more than medical science has revealed to us. It seems to be that we only seek to know more when things begin to break down and we become ill. Many that have sought help through using unorthodox alternative healing methods will testify to the reality of their existence and the powerful ways in which they can transform us.

Like all bodily organs and systems, the pineal gland only functions well if it is maintained in a healthy state. Its biggest problem is calcification. Fluoride, which in many geographical areas is added to drinking water, accumulates in and around the pineal more than it does with any other organ and leads to the formation of phosphate crystals. Over time, the gland hardens resulting in less melatonin being produced and subsequently the waking-sleeping cycle is disturbed. Perhaps this is why they say that as we get older, less sleep is needed. Other substances such as bromine and chlorine can also cause damage, so too can calcium supplements. One supplement that has a positive effect on the pineal though, as it does on the body itself, just happens to be a freely available natural source- the Sun. In fact sunlight is the best and only natural source of vitamin D, one of the most essential vitamins necessary for health.

There are also many other ways that we can both decalcify and maintain the overall functioning of the pineal, but rather than going into great detail here, I would suggest visiting an extremely good website; *http://decalcifypinealgland.com* where you will find some excellent information on the subject. In Chapter Seven of this book, I will look further at some of the more esoteric properties of the pineal that can be utilised to help connect with other frequencies and dimensions, as well as Dimethyltryptamine or DMT which is manufactured by the gland and is frequently referred to as 'the spirit molecule'.

DNA – the biological Internet

Another vital component of the 'physical' body that I have also written about before is DNA and it is so crucial to the story of light that I mention it again here. Located deep inside the nucleus of animal and plant cells are living chromosomes possessing a thread-like structure each containing a single molecule of DNA that functions like a holographic computer using 'endogenous laser radiation' - according to molecular biologist Dr. Pjotr Garjajev *(Fig 12)*. In an experiment Dr. Garjajev found a way to intercept communication from a DNA molecule in the form of ultraviolet photons - light! In addition he also claimed to have captured this communication from one organism,

Fig 12.

in this case a frog embryo, and then transmitted it, via a laser beam to the DNA of a different species (a salamander embryo) resulting in the latter developing into a frog. How is this possible? - *because both light and DNA are transmitters of information.*

In another experiment Russian scientists irradiated DNA samples, again with laser light and on a screen witnessed a typical wave formation forming. When the DNA was removed, the pattern remarkably remained in place and further investigation showed that the pattern continued to originate from the removed sample, where an energy field remained intact. The suggestion is that energy from 'beyond time and space' still flowed through 'activated wormholes' (energetic connections formed by the DNA) *even after its removal.*

Dr. Robin Kelly (1951-) whose books *The Human Antenna; Reading the Language of the Universe in the Songs of Our Cells* and *The Human Hologram; Living Your Life in Harmony With the Unified Field* both give great insight into the nature of DNA and its connection to the holographic universe, proposes that DNA acts like a transmitter/receiver of information from both physical and metaphysical realms. Some people have suggested that human DNA is a kind of *biological Internet* and I would agree with that. It communicates huge amounts of information in microcosmically small, but highly significant ways, mimicking a vast, possibly infinite network of information portals, rather like the billions of websites connected to each another across the globe. It may also account for abilities such as clairvoyance, intuition, healing and other forms of hyper-communication that mainstream science has yet to understand. Garjajev, it would seem, is of the same mind, claiming that organisms use 'light' to talk to each other.

Around 1970 Dr. Fritz-Albert Popp (1938 -) a theoretical biophysicist at the University of Marburg in Germany *(Fig 13)* was teaching radiology - the interaction of electromagnetic radiation in biological systems. In the time before mobile phones, with their accompanying cell-towers and numerous other systems such as WIFI which are now frequently linked with cancers and leukemia, Popp's studies centered around two similar molecules known to be harmful to humans; benzo[a]pyrene (known to be highly carcinogenic) and its almost identical twin benzo[e]pyrene. Having illuminated both molecules with ultraviolet light in an attempt to discover what made the two molecules so different Popp discovered that benzo[a]pyrene (the cancer causing one) absorbed the UV light and then re-emitted it at a completely different frequency. In effect it 'scrambled' the light and its coherent information. The twin molecule benzo[e]pyrene (harmless to humans) on the other hand, allowed the UV light to pass through it, unchanged. By repeating the experiment on other chemicals and compounds, some cancer causing, others not, he was soon able to predict which substances would actually be likely to cause cancer. In every instance, the ones that proved to be carcinogenic absorbed

Fig 13.

the UV light and then scrambled the frequency. Crucially, each of the carcinogens reacted only to light at the specific frequency of 380nm in the ultra-violet range.

Photorepair

It is a well-known phenomenon in biological experiments that by blasting a cell with UV light and thus obliterating 99% of it, including its DNA, it can be almost entirely repaired in a single day simply by illuminating the cell with the same wavelength at a much weaker density. Does this sound familiar to you homeopaths? Even though scientists still don't understand why this occurs, they cannot dispute that it happens and they refer to it as *photorepair*.

The significant leap that Popp made was to recognise that photorepair works most efficiently at 380nm, the identical frequency that the cancer causing compounds respond to and he reasoned that their reactions must in someway be linked. If this were true, he knew that it implied there must be some kind of light within the body responsible for photorepair and also that the compound responsible for causing cancer must be blocking and scrambling it. But how could it be proven?

Fortuitously, Popp came into contact with a student, Bernhard Ruth, who inquired if Popp would supervise his work. This he agreed to do if in return the student could show that light was emanating from the human body. As it happened, Ruth was a formidable experimental physicist and although he initially scoffed at the proposal, set to work to build the equipment necessary to test out Popp's theory, believing that he would prove his hypothesis wrong. Within a couple of years, Ruth had devised and built a device capable of 'counting' light, photon by photon.

By early 1976 they began trials on cucumber seedlings and Ruth's photomultiplier revealed that the fledgling plants were emitting photons, or light waves of a high intensity. To prove that this was not just an effect of photosynthesis, the scientists moved on to potatoes, deciding to grow the seedling plants in the dark. To their astonishment, when the seedlings were tested they showed an even greater intensity of light. Intrigued, Popp began thinking more deeply about the whole process of light within nature. He already knew that light was present in plants and that it featured in the process of photosynthesis and he reasoned that when we consume plant foods, we must also be taking in their light and storing it.

When for example, we eat certain vegetables and digest them, they are metabolised into carbon dioxide (CO2) and water, plus the light stored from the Sun and photosynthesis. We extract the CO2 and eliminate the water but what happens to the light? Surely, as an electromagnetic wave (EM) it must be stored? It appears that this is indeed the case and the photons dissipate and become distributed across the entire spectrum of EM frequencies.

This energy is the driving force for all the molecules in our body and even before any chemical reaction can occur at least one electron must be activated by a photon of a certain wavelength and sufficient energy. It also appears that the body *purposely* directs chemical reactions by means of electromagnetic vibrations, or what are termed 'biophotons'. All of this and much more occur because *light is information*. It is also *intelligent* and certain 'structures' within the physical body are geared to both respond and interact with it upon multiple levels, most notably DNA.

Photons - the intelligence within the cell

The conventional view of a cell is that it is like a bag of water containing molecules that, by chance, interact with each other following random collisions and that those that have certain complimentary shapes, like jigsaw pieces, lock together, allowing biochemical reactions to occur. Although this model has been somewhat refined over time, the concept persists to this day. But perhaps a better way of considering the whole process of cellular communication is to compare it to an orchestra. Every orchestra has its conductor and both the individual musicians and their conductor play to the score written by the composer. Although each player may possess great musical virtuosity, it is the job of the conductor to unite the whole orchestra, pulling each instrument into line to form the overall sound. If we were to think of the composer as being the *ultimate intelligence* or *creator*, the music would represent the *light*, the conductor and his movements would represent the *bodily intelligence* or *biophotons* and the orchestra members, *molecules within each individual cell*. So photons, 'switch on' the body's processes in the same way that a conductor brings each instrument into harmony, within the piece.

Popp discovered that molecules within the cells responded to certain frequencies and that the vibrational range of the photons elicited a variety of frequencies in other molecules of the body - *they communicated*. It seems that an individual molecule emits a unique electromagnetic field capable of both sensing and connecting with the field of another complimentary molecule. The scientific term for this 'dance' is known as electromagnetic *resonance*. In other words, the molecules send out specific frequencies of electromagnetic waves which not only enable them to 'see' and 'hear' each other, as both photon and phonon modes exist for electromagnetic waves, but also to influence each other at a distance, thus becoming drawn to each other if vibrating out of phase (in a complementary way). Returning to our musical theme, another way of looking at this phenomenon is to consider a pianist striking a tuning fork next to a piano and the particular piano string, when correctly tuned to the same frequency starting to sing back to the vibrating tuning fork. This occurs through energy passing from the tuning fork to the piano string and *vice versa* allowing the vibration to last much longer than if they were not resonating to the same frequency. Does the term 'as above, so below' ring any bells? Surely, if the universe itself is electrical in nature, allowing cosmic interaction to exist between galaxies, it is quite likely that the composition of other living organisms should be structured similarly, or am I missing something?

Interestingly, more evidence was emerging to corroborate what Popp had himself surmised. Irena Cosic, Professor of Biomedical Engineering, RMIT University, Melbourne, Australia, gained her Masters and PhD., and worked in Belgrade, Serbia, before moving to Australia in 2002. While still in Belgrade, she began as a graduate student of Dr. Veljko Veljkovic who now heads the Centre for Multidisciplinary Research and Engineering, Institute of Nuclear Sciences, Vinca. In a published paper with the grand title *'The Real Bioinformatics Revolution: Protiens and Nucleic Acids singing to one another?'* there is a wonderful statement that encapsulates beautifully, their discovery:

"There are about 100,000 chemical reactions happening in every cell each second. The chemical reaction can only happen if the molecule which is reacting, is excited by a photon…..Once the photon has excited a reaction, it returns to the field and is available for more reactions….We are swimming in an ocean of light."

Yet for Popp, the single most important question remained; where was the light coming from? If you are even the slightest bit computer literate, you will be familiar with compressed files, usually known as 'zipped' files. They are useful for storing or sending information across the web in a condensed form, as the recipient can then unzip them, thus decompressing and releasing the file into its original form. This essentially, is what happens when light, stored within DNA, is unzipped. DNA behaves like the master tuning fork of the body, striking a particular frequency to which certain molecules follow. More than this, Popp also believed that he had stumbled upon the missing link in current DNA theory - how a single cell can transform into a human being.

It became clear, through repeated experiments that DNA was the key component in finding the answers to so many fundamental questions and with biophoton emissions Popp believed he had opened up another dimension of understanding. His research showed that even weak light emissions were sufficient to orchestrate bodily repairs. In fact they had to be of a low intensity because they were happening at the quantum level. Significantly, it also appeared that the number of photons emitted were directly proportional to an organisms place in the evolutionary scale. Rudimentary plants and animals seemed to require more energy, corresponding to a very high frequency EM wave well within the visible range. Humans on the other hand only needed around one-tenth as much light at the same frequency.

Light – the giver of health

I recall an incident from many years ago when and friend and I were strolling in a park on a spring day. A little way in front of us was a young man, aged around twenty-five, walking quite quickly, as if he were late for an appointment. As we observed him we both noticed, quite independently, yellow light emerging from the souls of his feet. It was as if it was 'leaking' from his body. For a moment or two we were both stunned and wondered if we were the only ones to have witnessed it. Apparently no one else had and we were left trying to work out what it was we had seen. Our conclusion was that for some reason, the man in question was 'losing' light from his body, although quite why this should be we never discovered. Perhaps there was some kind of healing process taking place or maybe he just had an abundance of light!

If you have ever injured yourself, perhaps cutting a finger or some other part of your body, you will know that other than cleaning the wound and applying a plaster or bandage (assuming its not a deep cut) it somehow starts to repair itself without you needing to do too much more. The immune system is wonderful in taking charge of this process and has a built-in knowledge of what to do. Healthy cells begin to reproduce copies of themselves in response to signals they have received and the tear in the skin and surrounding tissue starts to get filled in. When the job is done to the satisfaction of the organising intelligence, work ceases and all that is left, is perhaps a small scar, acting as a reminder of the injury that took place. But how exactly does the immune system do this? What intelligence directs it?

In biophoton emissions, Popp believed he had found the answer. The process needed to coordinate and sustain the necessary repairs, requiring intelligent communication, could only occur in a holistic system under the directorship of a central controller. The players in the healing orchestra had to have a conductor capable of directing the music in a coherent way. The significance in this discovery, with regards to both curative and preventative medicine are profound, particularly in the case of cancer.

Coherent light

The incidence of cancer and other 'modern' diseases such as Type 2 Diabetes and high blood pressure are often attributed to lifestyle choices. This may well be so to an extent, but I have always felt that the causes of these types of conditions are more complex than the medical profession tells us. Exposure to certain chemicals and foodstuffs, combined with hereditary and environmental factors most certainly play their part, but something else is missing in the equation. If we turn this on its head again and consider a normal healthy body we will find that at a cellular level, light plays a huge part. Coherence is key to health and if this is lost or compromised in some way, illness ensues.

Research has shown that in cancer patients for instance, there is a loss of natural periodic rhythms and coherence. It is as if something has sabotaged their internal communications system - their wires have been cut and their messages scrambled. If you've ever heard the term 'his light just went out' you will understand why. Light holds the key to health and happiness and all biological systems, in one form or another, are designed to function this way. If the food that we consume, the water that we drink and the air that we breathe are contaminated, it is not just the biological composition of these things that destroys our wellbeing - it is also their ability to deliver coherent light for our DNA to then process and direct purposefully throughout our body. Based upon this it can be stated that *carcinogenic chemicals are destroyers of light and its information.*

By way of example; the practice of treating farm animals as mere fodder for human consumption and the attendant procedures of factory farming that cause so much pain and suffering to these poor creatures has a hugely detrimental effect on the 'end product' that consumers eat. It may come as no surprise to you to know that in tests, the healthiest foods were those that had the lowest and most coherent intensity of light, whilst disturbances in the system increased the production of photons.

This is because in a stressed state, the rate of biophoton emissions goes up. If we were to devise a phrase to define health we could state that it is 'a state of perfect subatomic communication' whilst ill health could be thought of (as in the words of the Led Zeppelin song) a 'communication breakdown'.

Biophoton therapy is not only an extremely vital tool in determining the health of a wide range of substances, it can also be applied to medical treatments and aid various healing processes throughout the body. I was 'informed' many years ago during a meditation, that light would one day be used to heal. This knowledge isn't new. It's just that as a race, humanity has forgotten why and how to use it. To my way of thinking, it's common sense.

So is it possible to have too much light? - of course. Ever tried to look directly at the Sun? - oh yes, I was warned about that wasn't I? Well, too much of a good thing is not actually good at all. Take an illness such as multiple sclerosis. MS is a neurological condition affecting around 100,000 people in the UK alone and is a state characterised by *too much order*, with sufferers actually taking in *too much light*. Strangely, it is more common in areas further away from the equator and is virtually unheard of in places like Malaysia or Ecuador, but relatively common in Britain, North America, Canada, Scandinavia, southern Australia and New Zealand. No one knows why this is, but is it suspected that environmental factors may play a part. You might be forgiven for thinking that those living nearer the equator would have more exposure to direct sunlight than those in the northern hemisphere, thus causing more incidences of the disease, but this is not so. Could it be that regular, natural sunlight is better for us than we are led to believe and that less of it, combined with exposure to man made artificial light actually triggers a process within the body that causes it to overcompensate? I have little doubt that this is true. Also, there is a growing amount of research that suggests that a lack of vitamin D could be a factor in causing MS - of perfect coherence as being the optimal state between chaos and order and too much cooperation having the effect of limiting or removing the choice of improvisation. It's as if the players in the orchestra become stifled, harmony is lost and the music suffers as a result. It has been suggested that in terms of diseases such as MS, sufferers are actually drowning in too much light. Exactly.

As with most things in life, balance is crucial. So yes, too much of a good thing can be bad for you and not enough, equally so. The human body has evolved, or I would suggest, been designed to operate to maximum effect when all of its systems are working in harmony and each frequency, from consciousness through to 'matter' communicates as one. Just as a finely tuned engine needs oil to run properly without seizing up, so light is the lubricant that drives everything living thing on planet Earth and beyond.

Incoherent light

Before moving on I should also mention sources of artificial light and their effects on the body. Light of the type emitted from tungsten filament lamps and ordinary fluorescent tubes can be described as incoherent, with frequent and random changes of phase between the photons. Fluorescent lamps, unlike the Sun, generate visible light by a *nonthermal* mechanism and the standard 'cool white' lamp has been designed to achieve maximum brightness for a given energy consumption. Since these 'strip lights' are the most widely used light sources in offices, factories and colleges, most people living in industrialised countries spend much of their waking hours bathed in light whose characteristics differ markedly from that of sunlight. It is known for instance, that strip lights reduce levels of the hormone melatonin, which controls the body clock and is linked to cancer. There are also many studies that suggest melatonin levels (and by proxy light exposures) control mood-related symptoms, such as those associated with depression - especially winter depression, also known as SAD or 'Seasonal Defective Disorder'.

In addition, instead of a glowing filament, modern so called 'low energy bulbs' utilise argon and mercury vapour within a spiral-shaped tube which, when the gas gets heated, produces ultraviolet light. This stimulates a fluorescent coating painted on the inside of the tube and as this coating absorbs energy, it emits light. Recent scientific evidence shows these specific rays are particularly damaging to human eyes and skin. Low-energy bulbs are also known to cause problems for people who have lupus, an auto-immune disorder that typically affects the skin, joints and internal organs and irritation caused by ultraviolet light worsens the rashes and exacerbates joint pain and fatigue associated with the disease. When no longer needed, there is also the danger of exposure to mercury and all modern low energy bulbs need to be disposed of safely. Of course, this doesn't always happen and they get thrown away with the rest of the rubbish, to the detriment of the planet and any life with which it comes into contact.

When we consider the enormous amounts of money and research that have been expended to 'save the planet' and in relation to light, to characterise and exploit its biological effects, it seems that very little has been done to protect people against potentially harmful living environments. It's as if, with the introduction of modern bulbs at the expense of the old-fashioned filament types that have virtually been withdrawn completely in some parts of the world, we are unwittingly serving as subjects in some kind of mass experiment on human health. I can only hope that this attitude changes before major issues develop because light is too important an agency to be exploited to the detriment of our wellbeing.

Living forms are as much a part of light as light is a part of them. The relationship is a symbiotic one. This is no more evident than in the functioning of our various bodily systems, some of which we have discussed briefly here. And as any student of metaphysics will tell you, a living form does not end at the periphery of what is commonly thought of as 'physical matter', but does in fact, extend way beyond the perimeter of that which can be registered by human vision. So let's venture further into the realms of electromagnetic energy fields - the fields of light that both interpenetrate and surround all living tissue.

Light: The Divine Intelligence

3

Deeper Fields

"There is a deeper layer of activity that has barely been touched upon. This deeper layer has to do with the energies that integrate the animated, living, homeostatic body. The day will come when they too will be catalogued and their laws understood."

Dr. Kate Klemer (1989 -)

Since ancient times we have seen paintings and images of different spiritual teachers across various traditions but one thing that is common among many of them is the depiction of a halo surrounding their head. Sometimes this field of light is shown to surround their entire body and clearly the impression is given that these individuals are saintly types, possessing of great wisdom and virtue. There is some truth in this, because the aura (derived from the Latin *aere*, meaning 'air' or 'gentle breeze'), also known as the *human energy field* or simply the *auric field*, appears to radiate ever more brightly around more spiritually pure individuals. Although I don't consider myself to be in this category I have had my 'aura' photographed using a digital camera linked to a laptop computer and the image appeared to show my body enveloped in an almost totally white energy. These types of photographs are quite widespread nowadays since increased access to more sophisticated computer software became available. Quite whether we can trust the validity of these images from a scientific standpoint is somewhat debatable, but the spread of their popularity does suggest that people are becoming more open to the idea that there is much more to this life than flesh and bone.

In the early twentieth century a number of doctors and scientists attempted to prove the existence of the human energy field. Amongst these were Walter Kilner (1847-1920) who published a work titled *The Human Atmosphere* in 1911 and Oscar Bagnall (1893) who carried on from where Kilner left off with the publication of *The origin and Properties of the Human Aura* in 1937. Before computers were invented, the photography of energy fields around living things was confined to a single unique method. Kirlian photography is a collection of photographic techniques used to capture the phenomenon of electrical coronal discharges and is named after Semyon Davidovich Kirlian (1898-1978) who, in 1939 accidentally discovered that if an object on a photographic plate is connected to a high-voltage source, an image is produced. Kirlian had gained a reputation as the best local resource for fixing electrical equipment, and was regularly called upon to repair the apparatus of scientists from laboratories in his area. He happened to witness a demonstration of a high-frequency electrotherapy device that had been developed by the French physicist Jacques-Arsène d'Arsonval (1851-1940) and noticed that there was a small flash of light between the machine's electrodes and the patient's skin. He wondered if it would be possible to photograph it, so experimenting with similar equipment, he replaced glass electrodes with metal substitutes to take photographs in visible light and despite incurring a severe electrical burn, was able to take a striking photograph of what appeared to be an energy discharge around his own hand. Knowing that he was onto something special, Kirlian developed his process and over time, noticed that images taken of living human subjects varied from person to person. He and his wife Valentina demonstrated that their pictures showed a life force or energy field reflecting the physical and emotional states of the subjects photographed and ascertained that these images could be used to diagnose illness. Shortly after this discovery in 1961, they published their first paper on the subject in the Russian Journal of Scientific and Applied Photography. Since then there have been literally thousands of Kirlian photographs published and below are just a small selection, showing the characteristic light patterns emanating from and interpenetrating the 'physical' form. *(Fig 14)*.

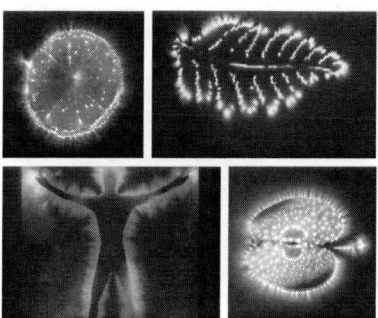

These images, as striking and as revealing as they are, represent only a fragment of the energy surrounding us and relative to humans, it can be said that there are several layers of energy representative of each corresponding frequency of being. Collectively, these represent our true auric field and form an invaluable part of our decoding mechanism. As I have revealed in my earlier books these fields allow continuous exchanges of *information and energy* relative to the frequency of each 'level' in accordance to the overall nature of the individual. As Nicola Tesla (1856-1943) famously said, *"If you want to find the secrets of the universe, think in terms of energy, frequency and vibration,"* and this is precisely what auric fields, through light, represent. American author, physicist and spiritual healer Barbara Ann Brennan (1939 -) writes of the human energy field being thus:

"The human energy field is composed of seven levels. Many people have the erroneous idea that this field is like the layers of an onion. It is not. Each level penetrates through the body and extends outward from the skin. Each successive level is of a 'higher frequency' or a 'higher octave.' Each extends out from the skin several inches farther than the one within it of lower frequency. The odd-numbered levels are structured fields of standing, scintillating light beams. The first, third, fifth, and seventh levels of this field are structured in a specific form. The even numbered levels - the second, fourth, and sixth - are filled with formless substance/energy. The second level is like gaseous substance, the fourth is fluid like, and the sixth is like the diffuse light around a candle flame. It is the unstructured level of the energy field that has been related to plasma and dubbed bioplasma.

Remember, these are not scientific terms we use here because experimentation has not yet proven what it is. But for lack of a better term, we shall use the word bioplasma. The bioplasma in all three of the unstructured levels is composed of various colours, apparent density, and intensity. This bioplasma flows along the lines of the structured levels. It correlates directly with our emotions. The combination of a standing light grid with bioplasma flowing through it holds the physical body together in its form, nurtures it with life energy, and serves as a communication and integration system that keeps the body functioning as a single organism. All of these levels of the human energy field act holographically to influence each other." - my thoughts exactly.

Another researcher who understood the principles underpinning living energy fields was Professor Harold Saxton Burr (1889-1973) whose books *The Nature of Man and the Meaning of Existence* and *Blueprint for Immortality* both reveal a deeper knowledge of the true nature of being. His studies of the electrodynamics of trees demonstrated both a connection and entrainment to earthly and lunar cycles and significantly he contended that the electro-dynamic fields of all living things, which can be measured and mapped with standard voltmeters, impress and control each organism's development, health, and mood. He referred to these as *'fields of life'* or *'L Fields'* and suggested that they are the basic blueprints of all life on this planet. I would concur with that and further suggest that these information fields and the light generated by them both inform and interact with DNA and the metaphysical universe. Burr wrote:

"Electro-dynamic fields are invisible and intangible; and it is hard to visualise them. But a crude analogy may help to show what the fields of life - L-fields for short - do and why they are so important. Most people who have taken high school science will remember that if iron filings are scattered on a card held over a magnet, they will arrange themselves in the pattern of the 'lines of force' of the magnet's field. And if the filings are thrown away and fresh ones scattered on the card, the new filings will assume the same pattern as the old. Something like this happens in the human body. Its molecules and cells are constantly being torn apart and rebuilt with fresh material from the food we eat. But, thanks to the controlling L-fields, the new molecules and cells are rebuilt as before and arrange themselves in the same pattern as the old ones."

"Until modern instruments revealed the existence of the controlling L-fields, biologists were at a loss to explain how our bodies 'keep in shape' through ceaseless metabolism and changes of material. Now the mystery has been solved: the electro-dynamic field of the body serves as a matrix or mould which preserves the 'shape' or arrangement of any material poured into it, however often the material may be changed."

"When a cook looks at a jelly mould, she knows the shape of the jelly she will turn out of it. In much the same way, inspection with instruments of an L-field in its initial stage can reveal the future 'shape' or arrangement of the materials it will mould. When the L-field in a frog's egg, for instance, is examined electrically, it is possible to show the future location of the frog's nervous system because the frog's L-field is the matrix which will determine the form which will develop from the egg."

Are you starting to make connections between the macrocosmic and microcosmic universes? Consider again the principle of 'as above, so below' and it can be clearly seen that both are electrical in nature. And why wouldn't they be? If the essential nature of the universe is electrical and it has an abundance of plasma energy, then why wouldn't bioplasma also be an intrinsic part of our make up? If the universe is filled with light, then why wouldn't we be also? If the nature of the universe is holographic, then why wouldn't we be too? If the universe contains dark matter, then why not *us?* If galaxies, stars and planets share information at non-physical levels then why wouldn't *we?* And here's the BIG question, which I've specifically worded in reverse - if we are sentient beings, and if consciousness exists within us, why wouldn't the universe itself be conscious?

Morphogenetic Fields – 'giving birth to form'

In the last chapter I briefly discussed the development of the 'physical' body in the womb and suggested that a non-physical holographic 'matrix' is formed at a higher vibrational waveform information level, unique to the individual. I further suggested that this could be considered as a kind of 'blueprint' or mould for the material form and that as the light of consciousness of the incarnating individual begins operating through the waveform matrix, decoding information from own innate higher self in combination with that already existing within the genetic formulation of the egg, so the formulation of the fetus develops.

It is my also contention that at the waveform level there resides information, stored holographically, to which individual consciousness is connected. This field is all encompassing, incorporating complete historical records relative to the creation of human form and imbibed with the intrinsic memory of all past, present and future events and is a crucial component, essential in the mechanics of all creation. Rupert Sheldrake (1942 -) developed the concept of *morphic resonance* which posits that *memory is inherent in nature* and that we inherit a collective memory from what has gone before. Some of Sheldrake's proposals have been characterised as pseudoscience, but I believe that his ideas have substance and my own research bears this out.

Fig 15. Rupert Sheldrake

In an email interview with the science journalist John Horgan, the author of *The End of Science* Sheldrake defined morphic resonance as:

"*....the influence of previous structures of activity on subsequent similar structures of activity organised by morphic fields. It enables memories to pass across both space and time from the past - the greater the similarity, the* *greater the influence of morphic resonance. What this means is that all self-organising systems, such as molecules, crystals, cells, plants, animals and animal societies, have a collective memory on which each individual draws and to which it contributes. In its most general sense this hypothesis implies that the so-called laws of nature are more like habits."*

It makes perfect sense to me that an individual consciousness or if you prefer the term, 'soul', about to incarnate would draw on it's own past energetic information (stored at the waveform level) and also the genetic blueprint of the race (also stored at the waveform level) and be able to decode the information when creating its new 'physical body'. In fact it would already, automatically be connected to its own unique morphic field, containing its past lives and history and also to the greater morphic field of the entire human race, encompassing the complete genetic and evolutionary history of the race and of the planet.

If perhaps like me, you have ever bought some flat pack furniture from one of those out-of-town retail park stores only to be totally overwhelmed when faced with assembling it, you'll know that the first thing you look for is the instruction sheet. This of course, is essential to every flat pack kit because without it the chances are that the wrong bits will be put in the wrong places and your bedside cabinet will end up looking like some strange piece of modern art. But because someone else has methodically worked it out for you, no doubt from bitter personal experience, their well written, illustrated instructions enable you to quickly work out where each part goes and before long your masterpiece emerges right there on the bedroom floor – job done. Everyone that buys the same flat pack has access to the same information and over time it might be that through experience, both the instructions and the design of the furniture itself are improved.

There exists enormous healing potential available through an understanding of how we are constructed and how we operate. There are plans for each body that we could consider as 'schematics' governing precisely the way in which they function across all levels - mentally, emotionally, physically and spiritually. Becoming aware of these can allow one to improve one's overall health and also strengthen the connection to higher levels in order to remove stress, fear, anxiety, confusion or pain. As an individual grows in awareness and the desire emerges to connect more deeply with the conscious intelligence or blueprint of each body, so alterations in frequencies can be made that impact on how that body and its energies operate.

Naturally, the question arises 'how exactly does the process work?' Sheldrake himself states:

"Morphogenetic fields work by imposing patterns on otherwise random or indeterminate patterns of activity. For example they cause microtubules to crystallise in one part of the cell rather than another, even though the subunits from which they are made are present throughout the cell."

Simply put, this suggests that the morphic fields to which an individual resonates - lets take humans as an example - impose their intelligence and crucially their information upon cells and I would suggest most critically DNA, through exchanges of light. Our DNA both receives and transmits information through light and the waveform fields to which we are connected capture this. Sheldrake also postulates:

"Morphogenetic fields are not fixed forever, but evolve. The fields of Afghan hounds and poodles have become different from those of their common ancestors, wolves. How are these fields inherited? I propose that that they are transmitted from past members of the species through a kind of non-local resonance, called morphic resonance. The fields organizing the activity of the nervous system are likewise inherited through morphic resonance, conveying a collective, instinctive memory. Each individual both draws upon and contributes to the collective memory of the species. This means that new patterns of behaviour can spread more rapidly than would otherwise be possible. For example, if rats of a particular breed learn a new trick in Harvard, then rats of that breed should be able to learn the same trick faster all over the world, say in Edinburgh and Melbourne. There is already evidence from laboratory experiments that this actually happens."

The waveform aspect itself carries no energy as such but strongly influences how the particle aspect behaves. From the biologist's viewpoint, the wave can be seen as the morphic field for the particle. Larger, more complex biological organisms such as humans, have morphic fields that are considerably more than the sum of their parts and both the field and the 'physical form' are intimately associated in the development of all subsequent similar forms. Sheldrake asserts that these fields are not diminished by the passage of 'time', since they carry no energy and that in a similar way to gravitational fields, only add to each other. Thus, every place is 'filled' with the morphogenetic fields of all past forms. Fields from similar forms may also 'overlap' and create a kind of composite field that is stronger although perhaps 'fuzzier' than the field from each individual form but over time the newer forms will gradually override the older ones and dilute their importance, thus giving the opportunity for evolution to occur. The implications of this theory are enormous, not least for the notion of where memory is stored and it seems to me an easy step to take in considering that both mind and memory may be located not only within, but also *externally* to the brain and are accessed by individual consciousness.

The Akashic Field

Another piece in the jigsaw of the non-locally interconnecting field that has many similar attributes to Sheldrake's Morphogentic Field and dates back to ancient times is the Akashic Field. *Akasha* is an ancient Sanskrit term that means 'sky' or 'aether' and refers to the most fundamental of the five elements of the cosmos - the others being *vata* (air), *ap* (water) *agni* (fire) and *prithivi* (earth). Akasha embraces the properties of all five elements and is the ocean from which everything emerges and unto which everything returns. It underlies and becomes all that manifests within the universe and is said to be so subtle that it cannot be perceived until it does so. Sound familiar? Yes, I thought it might.

A major advocate of the Akashic Field theory is author, system theorist and philosopher of science Ervin Lazlo (1932 -)

who has published many books and cutting edge articles that both challenge and redefine our understanding of reality including the excellent and groundbreaking *Science and the Akashic Field; An Intergral Theory of Everything*. Lazlo states that the world we observe is not the ultimate reality but an unobservable plane of reality lying beyond observed phenomena. He refers to *'the unified matrix'* - also referred to as the *Unified Field* by modern science, that harbours all the fields and forces, constants and entities that appear in spacetime, albeit being both beyond and prior to it.

Writing about the phenomenon of super-coherence in the universe Lazlo comments that it is *"responsible for all the phenomena in the world around us that we might describe as miraculous, awesome, or amazingly attuned."* He continues, citing the example of scientific experimentation carried out on living cells that demonstrates the non-local connection between them:

"Studies in macro-cellular biology have discovered that if you place two living heart cells next to each other - not touching, but just a short distance apart - they very quickly begin to beat in unison. They start off each contracting, pulsing, at their own rhythm, and very soon they are doing it at the same time, even though they aren't touching! But if there are disturbances in the environment, such as electric impulses or swirls of ionised water flowing around them, they won't sync up. So what is it that connects them? Clearly, they somehow tune to each other, but how? This is where the Akasha Dimension comes in. Since the Akasha Dimension is omnipresent and ever-present, it surrounds and flows through everything. And everything flows through it, but not just metaphorically. This means that things - stars, atoms, you and me - continually flow into existence through the information-binding

action of the Akasha Dimension that embeds all of existence by 'in-forming' - literally, putting form into - phenomena and events. What we do, who we are, and how we are is directly correlated with how well and how deeply we access the Akasha Dimension."

According to Lazlo the Akashic Field is a recognised scientific phenomenon, but quantum physicists refer to it as the Zero Point Field or ZPF.

The Zero Point Field

Quantum physics has revealed the presence of an all-pervasive background 'sea' of quantum energy in the universe. Dr. Harold Puthoff (1936 -) was one of the first to measure this energy at zero degrees Kelvin, the absolute lowest possible temperature in the universe, equal to -273 degrees Celsius. According to Newtonian physics at this extremely low temperature all movement of molecules and atoms should have ceased and no energy should be measured. What Puthoff found however, was what he called a 'seething cauldron' of energy and henceforth it was given the name zero point energy (ZPE). His discovery proved that the physical vacuum is not devoid of energy at all and that contrary to being a vacuum, space is actually a *plenum* (a space completely filled with matter).

The reality is that the ZPF is a consequence of something that particle physicists have long known about - the unpredictable nature of sub-atomic particles. The movement of these particles is believed to be directly caused by the Zero Point Field. Virtual photons randomly jump back and forth between the

ZPF and our 'physical' universe. They collide with and are absorbed by subatomic particles that are then excited into higher energy states. After nanoseconds the energy is released again by means of another virtual photon that returns to the ZPF. The photon is called a *virtual photon* because it comes and goes from the ZPF and does not appear to stay in the material universe. It is only used in the energy exchange between the two states.

It isn't only photons though, but all sorts of other elementary particles that emerge from the ZPF only to make an appearance for an extremely short time and then disappear again, back into the void. The Zero Point Field is a quantum foam of virtual particles and photons and our universe is never at rest, even at temperatures of absolute zero.

Now here is where it starts to get really interesting in relation to the story of light. A phenomenon known to science as Bose-Einstein Condensation - predicted as early as 1924 by Albert Einstein (1879-1955) and Satyendra Nath Bose (1894-1974), has far-reaching consequences. Bose had sent a paper to Einstein on the quantum statistics of light quanta (photons) that impressed him greatly. Einstein extended Bose's ideas and proposed that cooling atoms known as *bosons* to a very low temperature would cause them to fall (or 'condense') into the lowest accessible quantum state *resulting in a new form of matter.*

At the point where bosons occupy the lowest quantum state something known as *macroscopic quantum phenomena* occurs. This results in *superfluidity* - a state of matter in which the matter behaves like a fluid with zero viscosity. It appears to exhibit the ability to self-propel and travel in a way that defies the forces of gravity. Although this phenomenon was observed as early as 1938 when a substance known to science as Helium 4 was found to have zero viscosity (and no loss of energy) it wasn't until 1955 that the first Bose-Einstein Condensate was created in a laboratory at the University of Colorado, USA. In principle what occurs is that all of the particles within the condensate unify to become *one* - a single particle that is spread out across space and time. Furthermore, they draw their energy directly from the Zero Point Field in the form of zero point energy.

Interestingly, many of us are familiar with a Bose Einstein Condensate without knowing it. Whenever we play a DVD or CD the disk is read using a laser beam that is essentially composed of *coherent light*. A coherent light beam is composed of photons that have a constant phase difference and are all at the same frequency. Only a beam of laser light will not spread and diffuse. In lasers, waves are identical and in phase, which produces a beam of coherent light. Laser beams are also used to both create and read holograms and if, as I maintain, the universe is holographic in nature it can be seen that 'something' must exist to 'read' the information contained at all points within it. Logically, if we are decoding what we perceive as 'reality' from waveform fields into holographic form that same 'something' should also exist within each of us - *and it does*.

In the last chapter we discussed the functions of the pineal gland and suggested that it was a major component of the mechanism employed by consciousness to decode information. Well, here is more evidence of exactly how it functions; the human brain has direct access to the Akashic Field and the limitless amounts of information that it holds. If we draw upon an example that most of us are familiar with – cloud computing, then we can consider the Akash as being 'the cloud' in which all of our personal data and that of all others it remotely stored. When we connect with it, just as we would when using a mobile phone to either store or retrieve information, we effectively download portions of it for our personal use. Our mobile is designed and equipped with software to enable to information to be decoded and so too are we. Research has suggested that we generate huge quantities of coherent light within tiny structures located deep within the neurons of the brain. These are known as *microtubules* and they are so small that it is possible that the energy that they utilise to generate coherent light is zero point energy from the ZPF. Put another way, *this is a mechanism employed by consciousness* to decode and read information from the Akashic Record.

Another excellent researcher, physicist Bevan Reid detailed how 'virtual' energy of pure space converts initially into beams of biophotons, which, in turn, form the foundational matrix for living matter. Author Dr Robin Kelly, whose work I have drawn upon previously cited the work of Bevan himself in his own book, the aforementioned *The Human Hologram* stating;

'All space, including the proportionally vast areas within our atoms where no particles exist, comprises an infinite number of tiny spiral vortices, or wormholes, collectively known as the 'quantum foam.' Each one, according to this theory, acts as a portal to the entire library of universal information [Akashic Record] each one a mini star-gate to other dimensions. Reid proposed

that this space energy 'stored' within the vacuum of a symmetrical spiral makes contact with our dense living tissue, and then converts into the coherent laser-like beams I have already described. These beams of biophotons then contain both the infinite store of information from space, and also further information acquired from this collision with the physical hardware of our bodies. Our physical bodies are wonderfully efficient memory storage devices, complex computers containing the history of everything we have experienced. So it is proposed that space energy, on making contact with our bodies - our uniquely personal database of experiences - instantly downloads this vital information, storing it in its own vast files for posterity. There is therefore a constant and harmonious exchange of information between each of us and indeed all sentient beings, and the invisible field of universal consciousness. This process is known as resonance. So space energy is being continuously updated and upgraded by contact with our bodies, meticulously recording in its vaults all our earthly experiences. This record is enriched beyond measure by that special gift so valued by all human beings: the free will with which we are all so generously entrusted. So every deliberate act, every thought even, in every one of our unique lives contributes in a significant way to this universal, timeless field of consciousness. As each of us evolves spiritually, so in turn does the field. If we fully embrace this model, we human beings inherit an awesome level of responsibility. We are clearly, it would appear, co-creators of our own, our planet's, and our universe's destiny.'

These are powerful words indeed and ones that mirror my own understanding, but in order to explain more simply how all of this works, here is an illustration that I have adapted from an original idea proposed by Dr Kelly showing the steps involved in the process. *(Fig 16).*

Deeper Fields

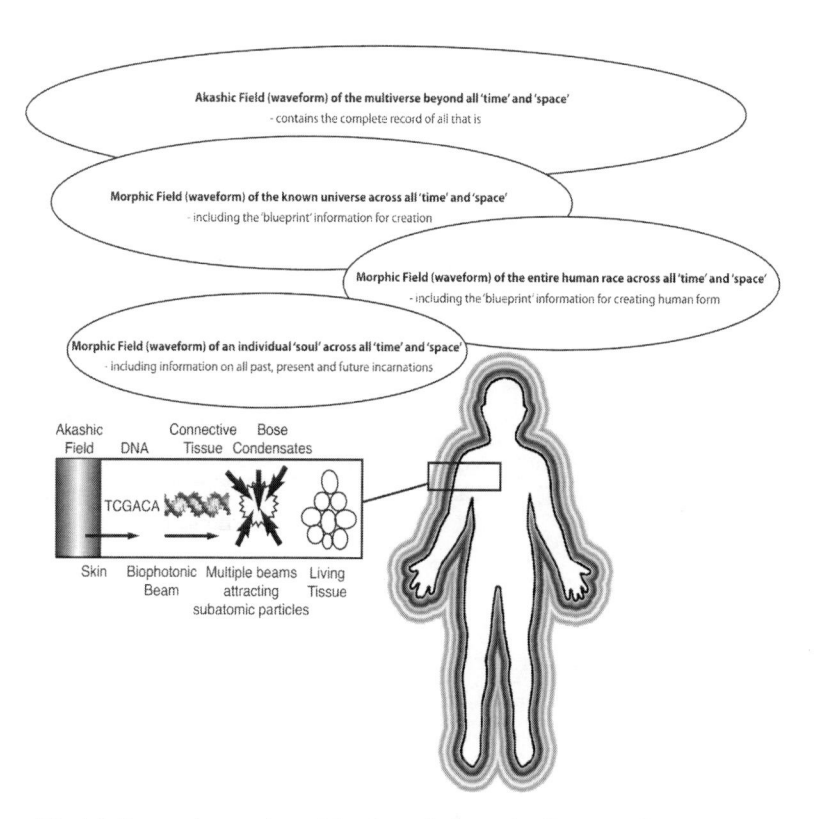

Fig 16. Bosons interacting with microtubules and cells - my adaption from the original illustration featured in The Human Hologram *by Dr Robin Kelly.*

The drawing of energy (and information) from the quantum field(s) via the Bose-Einstein Condensate has been likened to the way in which cold glass attracts water onto its surface but it is my contention that the process also works in reverse, allowing information to be uploaded via microtubules into the waveform fields with light playing a primary role. This may occur in several ways but there is further evidence that light is inextricably linked to the waveform information fields via a very strange natural phenomenon known as *sonoluminescense*. This is the transformation of sound waves into light energy and is well documented in mainstream physics. The effect was first discovered at the University of Cologne in 1934. Scientists put an ultrasound transducer in a tank of photographic developer fluid hoping to speed up the development process. Instead, they noticed tiny dots on the film after developing and realised that the bubbles in the fluid were emitting light with the ultrasound turned on. In another experiment a small spherical flask, filled with water was resonated by external sound speakers producing harmonious sound waves of 20 KHz. Then a very tiny air bubble was blown into the centre of the flask. When the air bubble was exactly centered in the vessel it started to rhythmically implode and emit light in ultra-short flashes representing trillion-fold concentrations of the original sound energy. Temperatures within the centre of the bubble reached an astronomical height of 30.000 degrees Celsius and created enormous pressure. Whilst mainstream physics is still having problems with this latter experiment and scientists continue speculating about cold fusion, others think that there is no other explanation than that the abundant light energy comes from the Zero Point Field - again, my thoughts exactly!

Blinded by the light

Just as a fish cannot know of the water in which it swims,

so we cannot see the air that we breathe or forces such as gravity that play such a vital part in our existence, although we most certainly experience their effects. On this basis it would seem quite feasible that we cannot 'see' the light of the Zero Point Field even though it exists. Whilst researching information for this book I 'accidentally' came across an article by respected German-born American astrophysicist Bernard Haisch whose works include *The God Theory* amongst others. In fact Haisch has published more than one hundred research papers on a variety of topics, many in prestigious journals and also served for ten years as an editor of the *Astrophysical Journal*. Reading through the article in question a paragraph jumped out to me immediately and knowing as I do that synchronicity plays a huge part in our lives I knew that I was meant to uncover the information. The statement that hit me right between the eyes was this:

"The solid, stable world of matter appears to be sustained at every instant by an underlying sea of quantum light."

If you've ever experienced a 'Eureka' moment - then this was mine. Here was a respected researcher who had reached the same conclusion that I had. I printed out the article and was excited further by what I read:

"The fact that the zero-point field is the lowest energy state makes it unobservable. *We see things by way of contrast. The eye works by letting light fall on the otherwise dark retina. But if the eye were filled with light, there would be no darkness to afford a contrast. The Zero*

-Point Field is such a blinding light. **Since it is everywhere, inside and outside of us, permeating every atom in our bodies, we are effectively blind to it.** *It blinds us to its presence. The world of light that we do see is all the rest of the light that is over and above the Zero-Point Field."*

This made perfect sense to me and if the penny hadn't already dropped, what Haisch had written confirmed my own thoughts exactly. Here is the rest of the piece - which I include in full because it works better that way. In each instance the bold emphasis is mine:

"We cannot eliminate the Zero-Point Field from our eyes, but it is possible to eliminate a little bit of it from the region between two metal plates. (Technically, this has to do with conditions the electromagnetic waves must satisfy on the plate boundaries.) A Dutch physicist, Hendrik Casimir, predicted in 1948 exactly how much of the Zero-Point Field would end up being excluded in the gap between the plates, and how this would generate a force, since there is then an overpressure on the outside of the plates. Casimir predicted the relation between the gap and the force very precisely. You can, however, only exclude a tiny fraction of the Zero-Point Field from the gap between the plates in this way. Counterintuitively, the closer the plates come together, the more of the Zero-Point Field gets excluded, but there is a limit to this process because plates are made up of atoms and you cannot make the gap between the plates smaller than the atoms that constitute the plates. This Casimir force has now been physically measured, and the results agree very well with his prediction.

The discovery that my colleague first made in 1992 also has to do with a force that the Zero-Point Field generates, which takes us back to $F=ma$, Newton's famous equation of motion. Newton - and all physicists since - have assumed that all matter possesses an innate mass, the M in Newton's equation. The mass of an object is a measure of its inertia, its resistance to acceleration, the A. The equation of motion, known as Newton's second law, states that if you apply a force, F, to an object you will get an acceleration, A - but the more mass, M, the object possesses, the less acceleration you will get for a given force. In other words, the force it takes to accelerate a hockey puck to a high speed will barely budge a car. For any given force, F, if M goes up, A goes down, and vice versa.

Why is this? What gave matter this property of possessing inertial mass? Physicists sometimes talk about a concept known as "Mach's Principle" but all that does is to establish a certain relationship between gravity and inertia. It doesn't really say how all material objects acquire mass. In fact, the work that Rueda, I and another colleague, Hal Puthoff, have since done indicate that mass is, in effect, an illusion. Matter resists acceleration not because it possesses some innate thing called mass, but because the Zero-Point Field exerts a force whenever acceleration takes place. To put it in somewhat metaphysical terms, **there exists a background sea of quantum light filling the universe, and that light generates a force that opposes acceleration when you push on any material object.** That is why matter seems to be the solid, stable stuff that we and our world are made of.

Saying this is one thing. Proving it scientifically is another. It took a year and a half of calculating and writing and thinking, over and over again, to refine both the ideas themselves and the presentation to the point of publication in a professional research journal. On an academic timescale this was actually pretty quick, and we were able to publish in what was *widely regarded as the world's leading physics journal, the Physical Review, in February 1994. To top it off, Science and Scientific American ran stories on our new inertia hypothesis. We waited for some reaction. Would other scientists prove us right or prove us wrong? Neither happened.*

At that point in my career I was already a fairly well-established scientist, being a principal investigator on NASA research grants, serving as an associate editor of the Astrophysical Journal, and having many dozens of publications in the parallel field of astrophysics. In retrospect, my experience should have warned me that we had ventured into dangerous theoretical waters, that we were going to be left on our own to sink or swim. Indeed, I would probably have taken the same wait-and-see attitude myself had I been on the outside looking in.

An alternative to having other scientists replicate your work and prove that you are right is to get the same result yourself using a completely different approach. I wrote a research proposal to NASA and Alfonso buried himself in new calculations. We got funding and we got results. In 1998, we published two new papers that again showed that the inertia of matter could be traced back to the Zero-Point Field. And not only was the approach in those papers completely different than in the 1994 paper, but the mathematics was simpler while the physics was more complete: a most desirable combination. What's more, the original analysis had used Newtonian classical physics; the new analysis used Einsteinian relativistic physics. As encouraged as I am, it is still too early to say whether history will prove us right or wrong. But if we are right, then "Let there be light" is indeed a very profound statement, as one might expect of its purported author. **The solid, stable world of matter appears to be sustained at every instant by an underlying sea of quantum light.**

But let's take this even one step further. If it is the underlying realm of light that is the fundamental reality propping up our physical universe, let us ask ourselves how the universe of space and time would appear from the perspective of a beam of light. The laws of relativity are clear on this point. If you could ride a beam of light as an observer, all of space would shrink to a point, and all of time would collapse to an instant. **In the reference frame of light, there is no space and time.** *If we look up at the Andromeda galaxy in the night sky, we see light that from our point of view took 2 million years to traverse that vast distance of space. But to a beam of light radiating from some star in the Andromeda galaxy, the transmission from its point of origin to our eye was instantaneous.*

There must be a deeper meaning in these physical facts, a deeper truth about the 'simultaneous interconnection of all things'. It beckons us forward in our search for a better, truer understanding of the nature of the universe, of the origins of space and time - those "illusions" that yet feel so real to us."

I'd like to add something crucially important to those fine words, the 'deeper meaning' that in my opinion Haisch refers to. It is simply this:

'Light is an expression of infinite consciousness'.

Haisch is correct when he writes that from the perspective of light all space and time collapses to an instant because there is no 'space and time' except for the somewhat limited consciousness of the observer experiencing life in a linear fashion through the lens of the mind. There is only *NOW*. There is only this moment and the light of consciousness is what gives everything in existence purpose and meaning.

Even one of the giants of quantum physics, David Bohm views the fundamental activity of existence as light and what he refers to as the *'holomovement'* (a dynamic wholeness-in-motion in which everything moves together in an interconnected process) is considered as an endless sea of light. In this paradigm matter is described as condensed or 'frozen' light and is composed not just of the visible portion of the electromagnetic spectrum but the entire range of waves that travel at the velocity of light. He also subscribes to the Big Bang (if indeed there was a Big Bang) as being conceived by light rays and suggests that even if the universe should ever come to an end, the universe of light, being beyond time, will have no end.

And why would it? Infinite consciousness, of which we are a part, had no beginning and will know no finality. It and we, are eternal.

Part Two
The Inner Light

Light: The Divine Intelligence

4

The Spectrum of Consciousness

"We are reflections of one another, therefore I know that you are part of me and I am part of you because we are all projections of the universal principles of creation/destruction, polarities of the same infinite consciousness that we call God."

David Icke (1952 -)

On a bitterly cold evening in November 1975 I was staying at a small but comfortable hotel situated on the seafront in Hastings, UK. I had travelled down by train from my then hometown of Birmingham for a weekend seminar organised by the Kent Healers Association. I was somewhat naïve for my age and at twenty-one this was my first solo trip away from home. As I introduced myself to the members of the group, who all seemed to already know each other, I began to feel distinctly uncomfortable and somewhat of an outsider. They were pleasant enough people, but I felt as though they were all so much more knowledgeable and spiritually advanced than I was and I anticipated a difficult weekend trying to fit in and make new friends.

Part of me wished that the ground would open up and swallow me.

Upon retiring to my small single bedded room for the evening I closed the heavy lined curtains and turned on my bedside lamp. The room was sparse except for the necessary furnishings and a small sink in the corner. After undressing and washing, I lay in bed pondering on the day's events. As I turned off the lamp the room instantly blackened and such was the density of the curtains that not even the tiniest speck of light from outside could enter in and it wasn't long before I drifted off to sleep.

I awoke in the early hours of the morning to find the room bathed in a pale blue glow. As I felt for my spectacles I became aware of what appeared to be a translucent ball of energy suspended about half a metre above the foot of my bed. Its luminosity was immense but not blinding and as I stared at it only half believing my eyes, it seemed for all the world that it was communicating with me. There were no words, only silence accompanied by absolute stillness and peace so I turned on the bedside lamp and flooded the room with artificial light but the orb was nowhere to be seen. Having pinched myself to see if I was dreaming I switched the lamp off again, only to find that the vision had reappeared, as if it had been there all the time.

The next thing I recalled was that morning had arrived and the room had returned to total blackness. It was only upon opening the curtains that daylight streamed in and I could see for the first time the shoreline that lay across the busy road. It wasn't though, until I walked down to breakfast that I realised fully how things had changed. People that had previously been distant and aloof were coming up to speak to me. Strangers that I hardly knew seemed like old friends and wanted to involve me in their conversations. I felt myself physically relax as a sea of calm washed over me and although I couldn't say for sure what had taken place, I knew that in some way events had transpired to make the weekend easier. Over the years I have often reflected upon the mysterious 'visitor' that came to me that night and like the famous character Scrooge in the Dickens book 'A Christmas Carol' have wondered if a similar fate befell me?

Left-brained science would I'm certain, have difficulty is ascribing the above to anything other than my own mind playing tricks in combination with my desire to find friendships in a difficult situation, but I believe otherwise. It is my belief that at a deeper level an exchange of information and energy took place between the light source and myself. I have no verifiable evidence that the orb possessed intelligence or that it communicated with me but there was a definite sense of presence that accompanied it. If you have ever felt that there is something else present in an empty room or had the experience of being stared at only to turn around and find another person doing precisely that, then you will know what I mean. Actually there is also a very interesting postscript to this little story in that the following morning after breakfast, I attended the first lecture of the day and as I took my seat awaiting the arrival of the speaker, I spotted an elderly lady staring at me from across the room. After a few minutes she came across and took the seat next to mine, opening her handbag as she did so. She then somewhat nervously introduced herself before handing me a crumpled piece of paper that she urged me to read. The lady said that she was staying in the room adjacent to mine on the first floor and that in the middle of the night 'someone' (from another dimension) had awakened her and impressed her to write. She went on to say that she was a medium who had the gift of 'automatic writing' and that those in the next dimension of life had a message for me. The note certainly seemed to bear out her words and the evidence she gave me that morning has stayed with me ever since.

It isn't the purpose of this book to provide evidence of whether individual consciousness persists after 'death' - there are many books that do that, including those written by psychics and spiritualist mediums as well as several compiled by my wife Amanda and I from our own extensive experiences. However, in the context of our study of light, there is much that is worthy of exploration, not least of which is *the light of consciousness* itself.

What is consciousness?

Ask a scientist that question and you may well find that the answer they give is that consciousness is an epiphenomenon of the brain - it 'kind-of' evolved quite by accident as a part of natural selection. In fairness, not all scientists believe this and an ever-increasing number of them now take the view that not only is consciousness the key component of life but it is actually responsible for its creation.

The difficulty that arises when attempting to define consciousness comes from the fact that it is not tangible or concrete. It is hard to define because it is another of those phenomena whose *effect* can be seen, but whose cause remains hidden. Over the years scientists and philosophers alike have come up with theoretical structures and concepts of what consciousness is and nearly always their ideas are bound up with 'mind'. It is easy to see why this should be so because all the evidence points to sentient beings having the capability to think and thought appears as a product of mind. Undertake a computer search for consciousness and you will find many definitions, some broader than others, incorporating the term 'mind'. A couple that don't but which are nevertheless incomplete in my view are the Oxford English Dictionary that describes consciousness as:

1) The state of being able to use your senses and mental powers to understand what is happening.
2) The state of being aware of something.

Whilst the Cambridge English Dictionary that has consciousness as being:

1) The state of understanding and realising something.
2) The state of being awake, thinking and knowing what is happening around you.

It is difficult to disagree in part with those definitions but neither actually define what consciousness is - they only describe what happens as a consequence of it. So we are faced with two main issues:

1. What is the origin and nature of consciousness?
2. What is its modus operandi?

For me, the answer to the first question is relatively simple - consciousness is the intrinsic essence of the divine, omnipotent creator, existing throughout and beyond all time and space. It is non-local and non-physical intelligence. It is light of the highest order. The second question is a little more difficult and gives rise to numerous further questions such as: How is 'self consciousness' different from 'consciousness'? What is 'unconsciousness'? What is 'super-consciousness'? At what level does consciousness emerge? Are you and I more conscious than a flower or a stone? Is the quantum field conscious and if so is it aware of itself? How does consciousness move matter? To coin a well-used phrase 'the list is endless.'

Through the ages, philosophers and learned thinkers have devised concepts to tackle these difficult questions. The idea of the unconscious for example, was introduced by Sigmund Freud (1856-1939) and further developed by Carl Jung (1875-1961). Freud's concept was that the unconscious was the warehouse for repressed drives and he used psychoanalysis in an attempt to bring these drives into conscious awareness. In contrast, Jung saw the unconscious as twofold - personal and connected to the vast collective unconscious of all humanity. Furthermore he saw it as transcendental, expressing itself in universal symbols - hence Jung's famous archetypes. Jung's theories about the creation and development of our universe incorporate a source from which all reality emerges and one of its main features is that each point, no matter how small, contains the whole. This base state he referred to as the *Pleroma* and the correlation between this idea and modern thinking about the holographic universe naturally come to mind. The three dimensional universe which sprang forth from Jung's Pleroma, he called the *Creatura* with the aforementioned archetypes being formative elements originating in the Pleroma. Jung expanded his theory to include the collective unconscious and suggested that the archetypes are patterns and symbols available to everyone, from all cultures past and present.

Jung's subsequent development of theories surrounding the *Ego* saw him viewing the self as transcendental - a higher directing force, surpassing that of the ego. His views surpassed earlier Freudian concepts with the suggestion that human reality emerges from a much deeper source. Others have taken on this idea and the suggestion of a hierarchic chain of consciousness that evolves from lower to higher states, eventually returning to the source is a popular one.

A working structure of consciousness, if such a thing exists, was expanded upon by Aldous Huxley (1894-1963) in his book *The Perennial Philosophy* published in 1945. Huxley wrote of the eternal self as 'being on a journey' with ever widening boundaries. In more recent times Ken Wilber (1949-) the American writer, philosopher and transpersonal psychologist who has extended his studies beyond psychology and has integrated aspects of Perennial Philosophy into his work has described six levels of consciousness *(Fig 18)*.

> **Ultimate** consciousness as such, the source and nature of all other levels.
>
> **Causal** formless radiance, perfect transcendence
>
> **Subtle** archetypal, trans-individual, intuitive
>
> **Mental** ego, logic, thinking
>
> **Biological** living (sentient) matter/energy
>
> **Physical** non living matter/energy

In this model each ascending level both transcends and includes all lower levels and because the higher transcends the lower it cannot be explained by or derived from the lower. In addition, although a higher level may contain attributes of a lower level it also has new aspects clearly distinguishable from those of the lower and therefore cannot be viewed as a derivation of the lower plane. A useful analogy of this principle would be to consider a two-dimensional world. A being living on this flat plane would have no awareness of a three- dimensional world but a being in the three-dimensional world would have awareness of the two-dimensional world. What seems to be speculation and imagination on one level becomes reality on another. Wilber further sub-divides each level of consciousness and these in turn, have rules and other implications, but I won't expand on those here.

The core principle of this model is one of movement and expansion and consciousness can perhaps be more easily understood when considered as a spectrum. Central to the Perennial Philosophy is the notion of the Great Chain of Being. According to the Perennial Philosophy, reality is not one-dimensional but is composed of several different but continuous dimensions. Manifest reality, consists of different grades or levels, reaching from the lowest and most dense (and least conscious) to the highest, most subtle and most conscious. At one end of this continuum of being or spectrum of consciousness is what we would refer to as 'matter' (or the insentient and the non-conscious) and at the other end is 'spirit' or 'Godhead'. Sometimes the Great Chain is presented as having just three major levels: matter, mind, and spirit but other versions refer to five levels: matter, body, mind, soul, and spirit. It doesn't really matter which of these, if any, you subscribe to. – it's your choice. The idea that an individual consciousness can grow and expand as it moves through the levels is central to the concept. What I haven't touched upon yet but which in my view is an integral part of this model, is *reincarnation*. If, as is suggested, consciousness experiences each successive level as it 'progresses' towards its ultimate goal of union (or re-union) with the Godhead then it must require consecutive vehicles (physical forms) through which to operate. It has to be able to draw energy and information from the waveform fields to which it is vibrationally connected in accordance with its level of progress in order to manifest those vehicles at each corresponding level.

I once came across a verse, attributed to the Kabbalah

that seemed to sum up the process very well:

> *God sleeps in the mineral,*
> *Dreams in the vegetable,*
> *Wakes in the animal,*
> *And becomes self-conscious in man.*

Rumi though, adapted this very nicely by changing a couple of words and I think I prefer his version:

> *Consciousness sleeps in minerals,*
> *Dreams in vegetables,*
> *Awakens in animals,*
> *And becomes self-aware in human beings.*

In *Transcognitive Spirituality* I established the concept of numerous levels or 'planes' through which individual consciousness experiences and interestingly my views are not too dissimilar to Wilber's interpretation of Huxley's work. I wrote that:

There are a hierarchy of realities through which life itself experiences and the illustration (Fig 19) shows that as consciousness expresses through form at the lower end of the evolutionary scale (1) its experience encompasses only the physical and etheric levels, with no awareness or development of the astral body. Pure infinite consciousness at these levels has not yet individualised and operates through simple forms at the very lowest end of matter, such as single electrons through to more complex forms like viruses. There may be many 'lives' within this frequency range before the opportunity presents itself to experience an expansion into more developed vehicles and this evolutionary track enables a hierarchy of more complex systems to develop through which consciousness can gain greater expression, what could be referred to as the implicit order of evolution.'

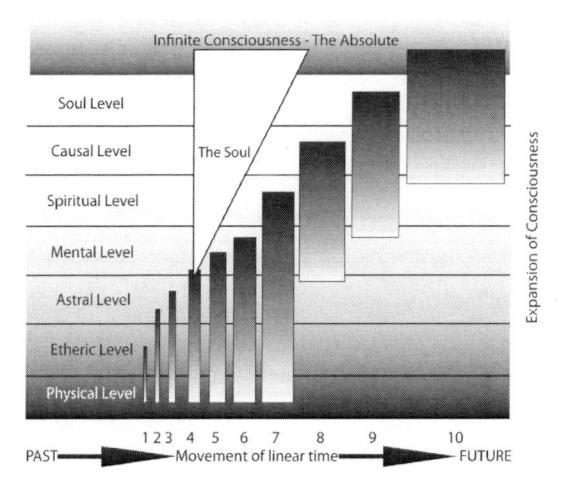

(Fig 19) The expansion of consciousness

My narrative continued:

'Moving above the etheric plane and into the next dimension of being, we see the emergence of the astral level. There is still no true individuality here as far as consciousness is concerned but more evolved forms and species appear allowing for an increasingly broader experience (2). At the lowest end of this level we find rocks and minerals and also elements such as water. More advanced minerals such as crystals also exist here and as we have already

discussed, water contains a form of memory that responds to words and emotions whilst crystals themselves have many properties that allow energies to be channelled or magnified with great effect. At the higher end of this level exists the plant kingdom and one of the main differences that we observe in comparison to the mineral reality is one of growth and reproduction. The emergence of a 'sex life' in terms of being able to reproduce (even with the help of insects) is a step up from the previous level and here the quality of consciousness is higher as is the response to stimuli. Plants as we have discovered also respond to the energy of thoughts and emotions in positive or negative ways.

Going further up the chain we reach the animal reality (3) and here we find that the astral reality really begins to exert a more powerful influence through a broad range of emotions. Movement becomes a choice as limited freewill is experienced for the first time, as is interaction across species with animals communicating with one another and in some instances with humans.

As within all levels there exists a hierarchy of forms and it can clearly be seen that certain species are more advanced that others. Insects for example - another of the forms found at this level, are able to make limited choices and decisions about their movement and behaviour but quite clearly they are not as intelligent as groups higher up the evolutionary ladder whose options are greater (4). This is because the emergence of 'mind' and the development of the brain within higher animal species enables consciousness to not only 'think' but more importantly reason, allowing for rational choices to be made that involve more than just instinctive behaviour. Whist some less developed creatures at the lower end of this range may operate through a group consciousness or be connected through operation of the aforementioned morphic field, the likes of humans and some of the more evolved species such as dolphins, whales, cats, dogs, apes, horses and others clearly operate more independently of each other even though connected at the most fundamental levels.'

At this point I introduced the emergence of the *'soul'* and expanded this aspect further by elaborating on the hierarchical structure of *'group souls'* and the way in which these further interacted with the 'physical' dimension both in terms of aspects reincarnating together to undertake particular tasks - including the paying back of Karmic debt and working with individuals still incarnate on earth engaged in important spiritual work for the furtherance of human understanding. The latter is particularly relevant in regard to mediumship and those that hold prominent and influential positions on earth. Doctors, surgeons and scientists, to name just a few, may never realise just how much they are influenced from higher, unseen dimensions.

Another author who has written extensively about the hierarchy of consciousness is David R. Hawkins MD, PhD (1927 -). He has created a tool that he calls 'The Map of Consciousness' (also called *The Scale of Consciousness*) with the ultimate goal of connecting the dots enabling the hidden picture to emerge. The core idea behind this is to remove the very sources of pain, suffering, and failure that hold us back and to assist the evolution of human consciousness to rise to the level of joy that should be the essence of everyone's experience. Hawkins developed the scale using a muscle-testing technique called *Applied Kinesiology* (AK) to document the nonlinear, spiritual realm. Based on the idea that the body does not lie, Kinesiology uses gentle muscle-monitoring techniques to access the subconscious mind and extract accurate information about current issues to restore balance to body, mind and spirit.

In Hawkins' chart each level of consciousness (LOC) coincides with determinable human behaviours and perceptions about life and represents a corresponding 'attractor field' of varying strength existing beyond our three-dimensional reality. There is a critical point within each LOC from which its field gravitates (or entrains). The numbers on the scale represent logarithmic calibrations of the levels of human consciousness and its corresponding level of reality but the numbers themselves are arbitrary – the real significance lies in the relationship of one number (or level) to another. There are some within the scientific fraternity who have been highly critical of Hawkins' work, suggesting (as is often the case) that it is pseudo-science. In my experience though, there are often deep truths to be found in areas that science has rejected simply because they fall outside of accepted scientific disciplines and are therefore worth further investigation. In Hawkins case some might classify him as a New Age teacher because of his opposition to disciplines such as channelling, fortune telling and Wicca. He also at times confusingly interchanges the terms truth, consciousness, vibration, frequency and energy on his calibration scale, bringing absolute objectivity into areas generally considered subjective and relative and according to his model there is little or no room for relativism - the concept that truth is not absolute, but relative to a particular frame of reference. Putting these issues aside for a moment, I still believe his work in worthy of consideration.

In my view, attempting to 'map' consciousness in a purely

objective way is a little bit like trying to count all of the grains of sand on a beach – almost impossible to accomplish, so perhaps what we should do is view Hawkins' map as a basic guide, something to be considered as a pointer – as one would a compass or an atlas and not definite proof. As a scholar I knew once said 'The map is not the landscape'. Hawkins ratings are as follows *(Fig 20)*.

Conscious Levels	Energy Level
Enlightenment	700-1000
Peace	600
Joy	540
Love (unconditional)	500
Reason	400
Acceptance	350
Willingness	310
Neutrality	250
Courage	200
Below	200
(Below the critical level of integrity):	
Pride	175
Anger	150
Desire	125
Fear	100
Grief	75
Apathy	50
Guilt	30
Shame	20

(c)1995 David R. Hawkins

These levels can be divided into positive energy-giving levels and negative energy-taking levels with the following traits and their consequences being characteristic of each, starting with the lowest and working upwards:

Shame - In Hawkins view, this is one step above death. At this level, the core emotion felt is humiliation. Those with suicidal thoughts and also those suffering from sexual abuse are often found here.

Guilt - One level above Shame, those residing at this level exhibit feelings of worthlessness, self-loathing and the inability to express self-forgiveness.

Apathy - This is the level of despair, numbness and hopelessness.

Grief - Although many of us have experienced this emotion at times of tragedy in our lives, having this as a primary level of consciousness can result in feelings of continual failure, regret and remorse.

Fear - Those living in abusive relationships may find themselves at this level. Also, those who feel persecuted and oppressed and who exhibit feelings of paranoia, suspicious of all around them. It may also include highly anxious people who 'perceive' the effects of fear as something they have no control over.

Desire - This is a major motivator for large swathes of human society. This is the level where addiction to ego, sex, money, prestige and perceived power is prevalent.

Anger - Moving up and out of Apathy to Grief and then Fear, individuals start to want. Any Desire that remains unfulfilled can lead to frustration and then to Anger. Subsequent recognition of Anger can be the catalyst enabling change but equally, lack of awareness can result in a prolonged stay at this level.

Pride - The halfway point and in Hawkins view, since the majority of people are apparently below this level, the one that most aspire to. In comparison to Shame and Guilt, feelings of positivity emerge here. However, this is a false state dependent upon external conditions such as position, wealth and power. This level is also the source of racism, nationalism, and of religious fanaticism.

Courage - The level of empowerment. It is the first level where individuals are not taking life energy from those around them. Courage is the recognition of self- empowerment and the realisation that the individual alone, is in charge of their personal growth and success. This is what makes people inherently human: the recognition of the gap between stimulus and response and the knowledge that there exists the potential to choose a response.

Neutrality - This is the level of flexibility. Being neutral implies that for the most part one is unattached to outcomes. At this level, there is satisfaction with the current life situation and the tendency not to need much motivation towards self-improvement or excellence in career situations. There may be recognition of possibilities but no desire to make the sacrifices necessary to reach higher levels.

Willingness - This is the level of the optimist. Those residing at this level see life as presenting them with endless opportunities and they strive to excel in everything they do.

Acceptance - Those at this level are goal setters and achievers, pushing themselves beyond their limitations and being proactive in their decision-making. Using the skills acquired from previously experienced stages, their inner potential is awakened. They have also recognised the spiritual wisdom in accepting some things as they are and going with the flow of life.

Reason - This is the level of the desire for knowledge, where the reasoning mind is more prominent and the need to know grows in strength. At this level, life's experiences are analysed into ideas and theories, wasting little time on things unlikely to be of educational value. One downside is that over-absorption in applying reason can result in an individual missing out on other expressions of consciousness.

Love - Here, if the lessons learned at the previous level are heeded is where intuition begins to take over from mind and where heart-centered thought and action takes precedence. Those exhibiting selfless love are in the minority but the ones that do are amongst most the most valued of all humanity.

Joy - This is the level of those considered spiritually advanced. As love becomes unconditional, the constant accompaniment of true happiness emerges. No personal tragedy or world event could ever shake someone residing at this level of consciousness. These individuals seem to lift up and inspire all who come into their orbit through being in complete harmony with infinite consciousness.

Peace - This is achieved after a life of complete surrender to the Divine Creator. It is where an individual has transcended all and reached a place of illumination, where stillness and silence of mind is achieved, allowing for constant revelation. Very few people on 'earth' ever reach this elevated level.

Enlightenment - This is the highest 'God-like' level of human consciousness, a state of perfect knowledge or wisdom, combined with infinite compassion.

Expanding consciousness and energy curves

The reason I have drawn upon Hawkins' Map of Consciousness model is that despite its limitations, it points toward an expanding personal awareness coupled with an increase in spiritual power. There is a definite evolution - a movement and growth from confinement to freedom, from ignorance to truth, from little to much and crucially from *darkness to light*. It is an upward curve, or as some might suggest an ascending spiral that is continuous and unbroken and which as it expands, results in ever-increasing levels of spiritual dynamism.

Conversely, lack of spiritual development in humans is characterised by lack of energy, lack of light and lack of choice. A poorly developed individual, defective in spiritual qualities, experiences limitations in freewill. Their personal options are curtailed, their perceptive abilities diminished and their intellect dimmed. It is as if their internal light has been turned off (or in some cases, not yet switched on) and they are compelled to fumble around in darkness until such time that something awakens deep within, allowing their higher consciousness to make a breakthrough. This can take a long time, perhaps across many 'physical' incarnations, *but it doesn't have to*. Life experience is a great teacher, but never let it be thought that spiritual enlightenment is something that must be accumulated over 'time'. Knowingness and wisdom can be realised in an instant. In fact, we are already *all-knowing* at higher, expanded states of being because *we are infinite consciousness*. It's just that we have temporarily forgotten (or been conditioned to forget) what we are and have fallen into the trap of believing what the mind-orientated sense of self feeds us through our limited five-sense perception. In fact, the world we think of as 'real' is largely, if not exclusively, oriented towards the mind and senses, severely inhibiting the expression of our greater self.

This aside, as the scales of ignorance begin to fall away and it starts to dawn on each of us what we truly are, so does the inner light of consciousness begin to awaken. In fact, it has never really been asleep, just unable to shine through the quagmire of mind-based beliefs and thoughts with which we are all too familiar. *The energy of consciousness is light* and this is the key component in the process of spiritual awareness. For the left brained, objective, scientific types who prefer formulae we could perhaps equate it as:

$$\text{Enlightenment} = \frac{\text{Consciousness / Light}}{\text{Awareness / Wisdom}}$$

I accept that this could be adapted to an extent to suit our personal view - we might for example wish to incorporate 'compassion', 'love' or other attributes into the equation, but for the purposes of this illustration I have assumed these to be already enfolded within my model.

I have previously spoken and written about what I refer to as *energy curves* and one way of thinking about these is to visualise an upturned cone with the sharp pointed end representing the minimal level of energy and the inverted base of the cone (which is infinite) representing the maximum level. At various stages in-between are varying points representing ever-increasing states of energy and expanding consciousness. This model *(Fig 21)* can be used to represent one individual across one 'lifetime', or across many 'lifetimes' and also expanded to reflect the total consciousness of either the human race or all sentient life upon earth - the possibilities are endless. Another more traditional model *(Fig 22)* sees the same theme adapted to incorporate the Alpha/Omega view, but still retaining the core principles.

Light: The Divine Intelligence

Fig 21.

Infinite levels of expanded consciousness

Enlightenment - **700 to 1000**

Ever-expanded consciousness

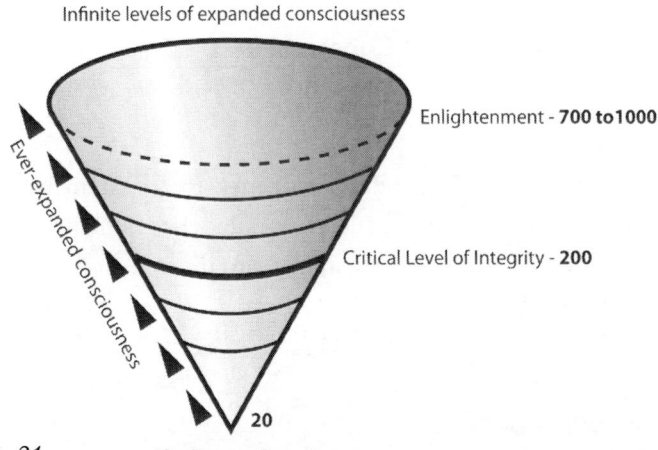

Critical Level of Integrity - **200**

20

The Energy Curve

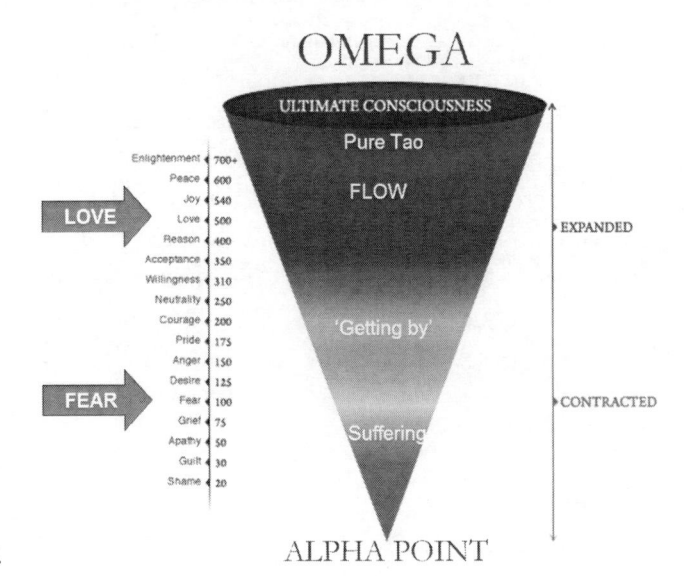

Fig 22.

OMEGA

ULTIMATE CONSCIOUSNESS

Pure Tao

FLOW

'Getting by'

Suffering

ALPHA POINT

Enlightenment	700+
Peace	600
Joy	540
Love	500
Reason	400
Acceptance	350
Willingness	310
Neutrality	250
Courage	200
Pride	175
Anger	150
Desire	125
Fear	100
Grief	75
Apathy	50
Guilt	30
Shame	20

LOVE →

FEAR →

► EXPANDED

► CONTRACTED

Coming into power

What does it really mean to have an expanded consciousness or a greater energy curve? Are these just New Age terms that are subjective and have no real meaning in daily life or do they refer to something much deeper and more tangible? In practical terms there may be no immediately obvious difference noted by someone with an energy frequency (according to Hawkins' chart) of between say, 200 and 350 (courage and acceptance respectively) but there would be a notable difference between an individual with an energy rating of 75 (grief) and 540 (joy). Just as the skies take a little time to clear after heavy and prolonged rain, it is unlikely that one would shift from grief to joy immediately, without experiencing the intermediate stages leading up to eventual transition. The same can be said in reverse, as witnessed when a person slides into depression, perhaps resulting from grief or a similarly lower emotional state, having previously been extremely happy and joyful. It is however, crucial to make the point that in someone with a truly expanded consciousness, increased inner light and expanded energy, the change is *permanent*. In our daily lives we can move through many levels of emotional shifts within a short space of time, but *real spiritual growth is irreversible*.

In computer terms, you can delete files and install new ones but at some level there is a record of all that the computer has undertaken, even if it is deep within the hard drive or somewhere in the 'cloud'. A new 'operating system' or programme though, will inevitably be more advanced and powerful than its predecessor, despite the fact that it may require more computing power and energy to operate in the way it was designed to, but the pay-off is that it offers the user more choices, more creativity and more power in a real sense. Although superior to earlier models whose operating systems offer less options and functionality, it is clear to see that without the predecessors, the advanced computer could not have emerged. In evolutionary terms, the one is dependant upon the other.

It is a matter of some conjecture as to what stage in the expansion of consciousness the accompanying augmentation of freewill and subsequent ability to impact upon the surrounding environment emerges. Some New Age teachers say that it is simply enough to 'ask the universe and it will be given'. But is it? From what I can see there is certainly a lot to be given (money, that is) from the courses they run to 'teach' people how to 'manifest their desires'. Whilst there might be some truth in the concept that we can all attract what we want in life, clearly it is not quite as simple as it is made out to be. There are all kinds of spiritual implications in respect of universal laws that influence the way in which our lives play out and the degree to which we can both access and apply our own unique freewill. At the frequency of infinite consciousness, which is pure light and unlimited potential, we may be all powerful, but at each successive level below that our light is diminished by degree and our choices curtailed. On this basis, individuals existing at the lower frequency range, have severely limited choices and have to rely upon artificially created 'power' and cooperation from others in order to get what they want from life. Often, those at these levels crave and sometimes attain positions of perceived power, becoming national leaders, dictators or despots by using those around them, whom they manipulate through fear or other lower emotional states to do their bidding, but their options are limited. In truth, all that they can manifest is darkness, through a severe lack of light within themselves. Notorious individuals such as Adolf Hitler, Pol Pot, Idi Amin and many others fit this description perfectly.

Someone with a developed mind, pure heart and expanded consciousness, whose thoughts, words and deeds are imbibed with wisdom and knowing, generates an inner light that extends way beyond what others would perceive as their physical being. This is not a light that can be seen in the visible sense, although even this is possible under some circumstances. Rather, it is the *inner light* and it is complimented by a host of magnificent attributes across a vast spiritual spectrum ranging from intuitiveness to the ability of conscious manifestation *(Fig 23)*. Just as our consciousness expands and we vibrate at ever increasing frequencies of being, so does our inner light increase and with it the accompanying gifts that enlightenment bestows. It is by no means possible to either describe or to illustrate the full extent of these attributes because they are largely subjective, inner states of being. Yes, a spiritually enlightened person may give off a radiant sense of calmness and peace and may appear outwardly serene and wise – both are external aspects of an inner state, but how is it possible to estimate or describe the powers they may possess? The inner light defies measurement by scientific criteria or definition by any spoken or written word. This is the point at which mind and intellect are left wanting and the light of infinite consciousness takes over.

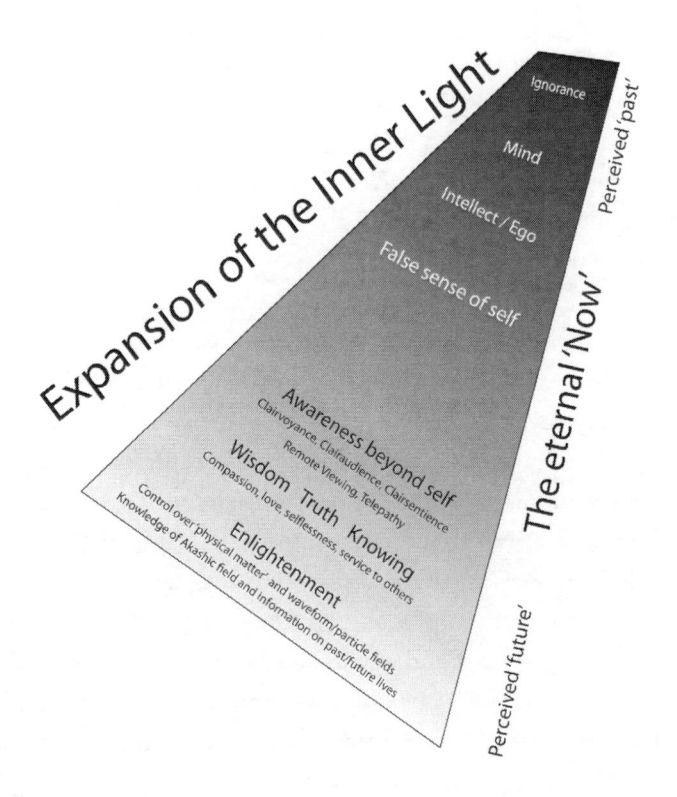

Fig 23. The expansion of the inner light is accompanied by ever increasing power and freewill choices, including the ability to manifest and manipulate matter.

Those who are on the path of spiritual unfoldment (all of us, really) are often drawn towards the light like moths to a flame, except that the light isn't external, but internal. We are drawn to the light of our own higher consciousness as it emerges through the darkness of our lower self. The main catalyst for this change, which almost exclusively begins at one of the lowest energy levels on the Hawkins' scale is through *suffering*. There are many causes of suffering and innumerable manifestations of pain, but all suffering is, I would suggest, of the mind – even that whose cause is perceived to be the result of physical trauma; inherited, 'accidental' or self perpetuated.

There are very few of us that pass through this world without any form of suffering. Yet suffering is not fruitless. Nor is it a punishment from any perceived God or higher deity, rather it is a consequence of the thoughts and deeds attributed to our own mind. Because the light of our higher consciousness is compelled to operate through the prism of the mind, it becomes splintered as its focus is directed through the lens of personality and the egoic self. Operating through the mental and emotional bodies, a form of spiritual amnesia ensues and the majority of people, at least in their formative years, forget what they truly are and from whence they came. Their belief is that they are a physical body with a gender, personality, cultural and national identity and limited life expectancy. From a physical perspective they are a body with a spirit (or even just a body) and seldom think of being *infinite consciousness having a human experience* (that it believes is real). There are exceptions of course, as some child prodigies demonstrate and it is not too difficult to spot an 'old soul' in a young body. Generally though, these individuals are in the minority and the majority of people, perhaps ninety-nine per cent of the population is, in degrees, ignorant of its origin and inherent infinite potential.

Suffering changes all of that. Often, if not always, it

provides the stimulus for an individual to question the reason and source of their pain and most importantly, to find *meaning* within it. Once this has been hinted at or identified, there emerges the impetus to escape the torment and move beyond the confines of suffering. In simple terms, when the 'lesson' has been learned there is no further reason for it to manifest. If one has been held within a darkened place and the light slowly begins to shine, there is no need to stay confined any longer.

One of the most wonderful examples of the emergence of the inner light can be found in the story of Victor Frankl (1905-1997), told in his famous book *Mans Search for Meaning*. During WWII Frankl was interned in Auschwitz, the notorious Nazi concentration camp where along with thousands of other inmates he suffered at the hands of his oppressors. After enduring the suffering in these camps, Frankl came to the conclusion that even in the most difficult, painful, and dehumanised of situations, life has potential meaning and that as a consequence, even suffering is meaningful. There are many superb quotes attributed to Frankl but this short extract taken from his account of labouring under the brutal conditions imposed on him by the Nazis encapsulates perfectly the insight into the human condition that he discovered:

We stumbled on in the darkness, over big stones and through large puddles, along the one road leading from the camp. The accompanying guards kept shouting at us and driving us with the butts of their rifles. Anyone with very sore feet supported himself on his neighbour's arm. Hardly a word was spoken; the icy wind did not encourage talk. Hiding his mouth behind his upturned collar, the man marching next to me whispered suddenly: "If our wives could see us now! I do hope they are better off in their camps and don't know what is happening to us."

That brought thoughts of my own wife to mind. And as we stumbled on for miles, slipping on icy spots, supporting each other time and again, dragging one another up and onward, nothing was said, but we both knew: each of us was thinking of his wife. Occasionally I looked at the sky, where the stars were fading and the pink light of the morning was beginning to spread behind a dark bank of clouds. But my mind clung to my wife's image, imagining it with an uncanny acuteness. I heard her answering me, saw her smile, her frank and encouraging look. Real or not, her look was then more luminous than the sun which was beginning to rise.

A thought transfixed me: for the first time in my life I saw the truth as it is set into song by so many poets, proclaimed as the final wisdom by so many thinkers. The truth – that love is the ultimate and the highest goal to which Man can aspire. Then I grasped the meaning of the greatest secret that human poetry and human thought and belief have to impart: The salvation of Man is through love and in love. I understood how a man who has nothing left in this world still may know bliss, be it only for a brief moment, in the contemplation of his beloved. In a position of utter desolation, when Man cannot express himself in positive action, when his only achievement may consist in enduring his sufferings in the right way - an honourable way - in such a position Man can, through loving contemplation of the image he carries of his beloved, achieve fulfilment. For the first time in my life I was able to understand the meaning of the words, 'The angels are lost in perpetual contemplation of an infinite glory.'

Each of us must discover the same truth that Frankl did, in our own way. It cannot be taught, only experienced, but when found, will never again be forgotten. Interestingly, having survived the horrors of Auschwitz, Frankl continued to develop his concept of *Logotherapy* which is founded upon the belief that it is the striving to find a meaning in one's life that is our primary, most powerful motivating and driving force.

Over the course of my life I have encountered many people whose path towards the goal of spiritual understanding and the discovery of meaning has taken them through suffering, leading to self discovery and the opening up of their spiritual reservoir of compassion towards others. My wife for example, has since a child undergone many painful operations as a result of being born with a condition known as 'Bi-Lateral, de-rotational congenital dislocation of the hips' that was missed at birth and only discovered when she reached the age of three. She has endured physical, mental and emotional trauma as a result of this condition, yet when I see her reaching out to others who are in the midst of suffering I never fail to be moved by the love and compassion that she exudes. I know without a doubt, that her life experiences have been instrumental in releasing the inner light that now radiates forth from her, to others.

In my own case the suffering I endured as a child was entirely of the mind. I was fortunate to have a happy childhood with loving parents, but from the age of around six until my early teens I was a hypochondriac. As an only child I enjoyed my own company, but became somewhat introverted. In some ways, this encouraged my quieter, reflective nature to emerge, allowing me to develop the ability to think deeply about life, but with the downside that I also had more time to focus upon my perceived ailments. I suppose I was a born worrier, although looking back now I can see that I must have picked up many negative traits from some of those around me. It wasn't until my late teens when I was given a book titled *Psycho Cybernetics* by Maxwell Maltz (1889-1975) that I began recognising that I was the architect of my own pain, no one else. By this time though, the suffering that I had endured had been sufficient to awaken my compassion towards others and the desire to serve in whatever capacity I could. My subsequent development as a trance medium and the spiritual work I have undertaken since is testimony to the inner light that emerged at that time and which has continued to illuminate my life to this day. I will leave it to others to judge the value of my service, but if it is considered that I have been a catalyst in helping others through my own example, then I will be thankful.

I must emphasise here, that whilst suffering can be transformative in opening the doorway to the inner light, it is not a specific requirement that all must endure. There are other experiences that lead to expanded states of consciousness and enlightenment, but always, without exception, true spiritual development is accompanied by humility, compassion, empathy, selflessness, desire for service to others and above all, love. These attributes are 'hard-wired' into the machinery of life and even though some of our human software may deflect our aim from time to time, eventually we come back on track with renewed purpose. I like to think of life's journey as being like that of a river, whose source is forever constant, pure and undiluted. Yet the descent into that which we decode as physical form presents obstacles and barriers to be circumnavigated along with challenges to be met, some of which colour and taint us. The further we appear to venture from the source, although it still flows within us, the more that these impurities discolour our thoughts and actions. For some, this results in the arising of many critical issues from which suffering ensues, before eventually the water begins to run clear again as it approaches the great ocean - the true origin of its being.

For others, their original clarity is largely retained, even across many 'lifetimes' and suffering is a comparatively rare experience, adding little to their spiritual expansion. Yes, they will experience it from time to time, but not necessarily as the catalyst for further growth required by so many.

A teacher once said 'Many pathways lead to one place' and this is so true. For each, the journey is unique, but the place at which all arrive is the same. The wellspring from which our stream of consciousness emerges is pure light and having passed through degrees of darkness, it is unto this light that we 'return', having never really left it in the first place. Self-realisation is the knowing that there is no 'self' - *all is one*. It is within this oneness that the gifts of the inner light, the pure eternal light, manifest.

5

Gifts of the Light

*"Soon, oh soon the light, pass within and soothe this endless night,
And wait here for you, our reason to be here."*

from 'Soon' by Jon Anderson (1944 -)

You will recall that in Chapter Three we commented upon David Bohm's suggestion that 'the universe consists of frozen light' and as recently as 2014 researchers at Princeton University, USA have been turning light into crystals - essentially creating 'solid light'. To them this was a 'new behaviour' of light, not seen before. For generations, physics students have been taught that photons, the subatomic particles that make up light, don't interact with each other. Yet the researchers were able to make photons do exactly that. One of the scientists exclaimed *"These interactions then lead to completely new collective behavior for light - akin to the phases of matter, like liquids and crystals, studied in condensed matter physics."*

What the scientists at Princeton are addressing here is the manipulation of light through physical means – using scientific apparatus to create a previously unseen form of light. This is entirely feasible because light has many properties and forms that we have yet to recognise or understand. But what is of even greater significance is our human potential for empowerment by the inner light, once we have reached a stage in our spiritual awakening that allows it. It is not an exaggeration to state that the fully enlightened being has the ability to manipulate light and matter (same thing) in the most remarkable ways.

Looking back through history there are numerous examples of highly evolved individuals or *'Avatars'* that have demonstrated what we would consider to be supernatural abilities. Whether or not you believe that Jesus was an actual person who lived over 2000 years ago (I personally don't) some of the 'miraculous' happenings attributed to him certainly fall into this category. Even if the man and his miracles were fictitious, others have also demonstrated many of the marvels attributed to him, including; healing, prophesy, hearing discarnate voices, communion with deceased people, resurrection after 'death', manipulation of matter, clairvoyance, mind-reading and telepathy to name a few. Indeed, I too have either witnessed or personally demonstrated some of these phenomena and so can testify to their existence.

A more famous example of someone who could be considered an Avatar is Paramahansa Yogananda (1893-1952) who was the first yoga master of India to take up permanent residence in the West. Although possessing of his own powers, in his famous book *Autobiography of a Yogi* the sage details many examples of spiritual masters personally known to him, who had the ability to demonstrate extraordinary feats. Yoganada's main teacher or guru was Sri Yukteswar (1855-1936) who was himself a disciple of Lahiri Mahasaya (1828-1895) – both of whom were self-realised spiritual masters. Lahiri Mahasaya's divine exhibition of miracles was seemingly endless and although possessing a physical body he was also formless (as are we all at higher levels), materialising in more than one place at the same time. Yukteswar too, was known for his spiritual wisdom and ability to demonstrate extraordinary feats. It is said that at one time he had written a spiritual commentary on the Bible that had not been published. He apparently spoke about it to a French gentleman who held an official position in the administration of the French settlement of Chandernagore. This man was very much impressed by what he heard and wanted to read the manuscripts himself. Accordingly the manuscripts were handed over to him and when the two men met again the Frenchman said that he had been overwhelmed by what he had read. He proposed that he might be permitted to take the manuscripts to Metropolitan France where he would show them to the scholars there, promising to bring them back on his return. Sri Yukteshvar agreed but never did get his work back as the Frenchman never returned to India and the invaluable manuscripts were lost. Nobody knows what the commentaries contained as Sri Yukteshvar did not discuss them with anybody. Perhaps the information only resides now within the timeless field of the Akash, something we may only discover when we leave this frequency range.

Light: The Divine Intelligence

Paramahansa Yogananda *(Fig 24)* also extolled the abilities of another sage named Gandha Baba who was able to make a person's skin exude the perfume of any chosen flower and writes:

Fig 24.

"*I was a few feet away from Gandha Baba; no one else was near enough to contact my body. I extended my hand, which the yogi did not touch.*

"*What perfume do you want?*"

"*Rose.*"

"*Be it so.*"

To my great surprise, the charming fragrance of rose was wafted strongly from the center of my palm. I smilingly took a large white scentless flower from a near-by vase.

"*Can this odorless blossom be permeated with jasmine?*"

"*Be it so.*"

A jasmine fragrance instantly shot from the petals. I thanked the wonder-worker and seated myself by one of his students. He informed me that Gandha Baba, whose proper name was Vishudhananda, had learned many astonishing yoga secrets from a master in Tibet. The Tibetan yogi, I was assured, had attained the age of over a thousand years.

"*His disciple Gandha Baba does not always perform his perfume-feats in the simple verbal manner you have just witnessed.*" *The student spoke with obvious pride in his master.* "*His procedure differs widely, to accord with diversity in temperaments. He is marvelous! Many members of the Calcutta intelligentsia are among his followers.*"

I too have witnessed this type of phenomena. Some years ago I attended a public demonstration by 'psychic surgeon' Stephen Turoff in Droitwich, UK. Prior to starting his surgery demonstration he gave a brief example of his clairvoyant powers before asking if someone would fill a bowl full of tap water. A member of the audience, which numbered around two hundred, dutifully complied and handed the bowl to Stephen. He asked if anyone would care to step forward and smell the water to make certain that it was not scented in any way before proceeding to walk amongst those present with bowl in hand. Stopping periodically to ask what scent each person desired, he then dipped his hands in the water and upon withdrawing them, enquired of the individual if they could smell their chosen scent on his hands, which they could, without exception. When he approached me asking what perfume I desired him to manifest, I replied 'lavender' and having dipped his hands in the water he asked me to clasp mine together before enclosing them in a gentle embrace.

"Now smell your hands," he said.

"They smell of lavender," I pronounced.

I watched as he continued to move along the line, never stopping to dry his hands as he passed from one person to another, always getting a successful response even though the requests were diverse with no two people asking for the same scent to manifest. Although some might suggest that this was a magic trick of some kind, I felt at the time it was genuine and if it was an illusion, then it was a very good one. Interestingly it states on Turoff's website that:

'One of his abilities is to 'Bring down the Light' which is used for healing as well as for enhancing spiritual development.'

On another page headed 'The healing power of light' it is written:

"Wherever he appeared, a special kind of light - the light of unconditional love with which he heals us - filled the room and lit up the people around him," say those who received a complementary treatment in July in Slovenia.
Some people tried to take photos of that light. Some of the most unusual photos were sent to Breda Vesel from Bled, which organizes the visits of Stephen in Slovenia.
The first recordings were a few hours after his arrival in Slovenia in the afternoon of the 5th July during his visit with the family in Verlic Šmartno at the foot of Šmarna gora taken near Ljubljana.
Stephen Turoff blessed the house and the grounds of the owners at their request. They had both lost their jobs and lived since then in accordance with the principles of self-sufficiency as described in the book about an unusual Russian-Siberian woman, Anastasia.
At first glance, it looks as if the photos on which he gives his blessing, were overexposed. The trees are not green. They are shrouded in a kind of light. "In the first picture you can see even before his speech a little of his light. But when he speaks to his audience, the light begins to spread out in all of us." Said Vesel Breda.

Later in the demonstration, Turoff (who was by this time under the spirit control of Dr. Kahn) performed 'psychic surgery' upon several members of the audience who sought healing for a range of conditions. By far the most powerful of these 'operations' was one performed on a gentleman who said that he had a problem with his eye. Turoff/Khan 'examined' him and told the audience that he had a tumour behind the eye that needed to be removed. What followed was nothing short of phenomenal as a white handkerchief was placed over the eye and the operation commenced. Lying prone and face upwards, the man did not flinch as the healer's hand rested gently upon the handkerchief for several seconds. After removing his hand Turoff/Khan stated that the eye had been 'dematerialised' and proceeded to gently push a finger into the empty eye socket. There were gasps from the audience as the handkerchief was seen to pucker, crumpling gently into the hole where only seconds before was an eye. As the surgery continued Turoff/Khan was seen to use a small unsterilised scalpel, similar to those used by graphic artists, that appeared to cut something out of the base of the socket. Used swabs of cotton wool were discarded along with some 'waste tissue' into a small bin, each accompanied by a gentle thud that was audible to all present. Upon completing the removal of the tumour the handkerchief was again placed over the empty eye socket and at this point a member of the public who, I later discovered was a conventional surgeon, was asked to step closer to the table as the man's eye re-materialised. As this occurred, the handkerchief appeared to fill out and push upward from the socket as the eye returned. The stunning demonstration concluded with the patient sitting up and after a few minutes standing and then walking from the table. At no point was any blood in evidence and it did not appear that the man in question experienced any discomfort at all. Also, upon examination, the waste bin contained only crumpled bits of cotton wool, nothing else. When questioned on this later, Turoff said that any tissue removed from the body of a patient also dematerialises.

Some friends who accompanied me to the event managed to speak to the mainstream surgeon who had witnessed the operation close up and he shook his head in amazement, confessing that "I could not do what he (Turoff) did this evening". Neither could he offer an explanation, other than it was a 'miracle'.

I encountered Stephen Turoff *(Fig.25)* again a couple of years later when I took a close friend to see him at his Danbury clinic. Although still channelling Dr Khan, the ambience of the small portacabin from which he operated had the definite feel and appearance of an ashram, with subtle aromas and vibuti (sacred ash) in evidence. This time I was able to stand only a few feet away from the operating table as Turoff, again deeply overshadowed by Dr Khan, made an 'incision' in the stomach area of my companion (although the blade of the knife at no time touched her skin) and removed something, I know not what, discarding it into the bin. At one point Dr Khan looked up at me and enquired if my own back was alright. I replied that it was, upon which he commented that I should be careful, as there was a weakness there - something that I can confirm, because I started to suffer some lower back pain only a few months later.

Following the fifteen-minute 'operation' my companion and I headed home and I later asked if she had felt anything during the surgery. "It was as if his hand was inside my body" she replied, adding that there was no pain at all.

I enquired if there was any scarring to be seen, but all that was in evidence was a small reddish line, about a millimetre in length, where the body had been 'opened up'. This vanished within a day or so. I won't disclose the complaint that my friend had prior to visiting the clinic, but to my knowledge it was healed and did not return. Stephen continues to undertake his healing work to this day and his website *www.stephenturoff.com* is well worth a visit.

Fig. 25

Naturally the question arises that if the there is no contact between the scalpel and the actual 'physical' body, what 'body' is it that is being operated upon? Current esoteric thinking suggests it is the 'etheric' body – a higher frequency body of which the physical is a duplicate. This may well be true, but equally it might be that the repair work is carried out at a higher waveform level, before being decoded by the individual's own consciousness back into the holographic physical reality that we experience.

Another modern day healer operating in a similar way to Turoff is João de Deus (1942 -), also known as John of God - the miracle man of Brazil. A website covering his activities proudly proclaims:

'He is arguably the most powerful unconscious medium alive today and possibly the best-known healer of the past 2000 years.'

Whether that is true or not some of those who have witnessed or been a recipient of his healing have testified to the tremendous power operating through him. Referring to an 'operation' on a gentleman, the website script, obviously written by an onlooker, details the healer's procedure as follows:

'John of God is scraping the surface of a man's eye. Again, this allows him to successfully treat a wide range of conditions throughout the body. The most probable mechanism I can offer is that the iris contains reflex points to all other parts of the body (see a textbook on Iridology). The man was awake and in no discomfort.'

Sounds familiar doesn't it? Another short paragraph states:

'John of God has inserted a surgical instrument up the nose of the patient. I can't explain the mechanism by which it heals illnesses throughout the body. But I can say, one of my friends had this operation and afterwards her comments were that it tickled but didn't hurt and all she wanted was for him to do the other side of her nose.'

Clearly, something extraordinary is taking place here and short of trickery, it seems that John of God, who like Turoff has mediumistic ability, is working on a non-physical level to perform the healing. Indeed, as Turoff surrenders a portion of his consciousness to Dr. Khan, so John of God purports to have more than one discarnate entity working though him. One of these is Dr Oswaldo Cruz (1872-1917). Research has shown that whilst incarnate Dr. Cruz was instrumental in treating patients with vaccines and critically instituting the modern medical practices of notification, quarantine of cases, eradication of pests carrying the disease and improvement to public hygiene. He died of kidney failure in 1917 and is remembered in Brazil as the father of sanitation and environmental medicine.

Another famous healer, also a medium, is Ray Brown. Born in 1946 Ray grew up in Portsmouth, UK. He later entered the building trade and became site manager for a construction company before training to become a healing medium having earlier experienced OBE's (out of body experiences) when he was as young as five years old.

At the age of seventeen he attended a demonstration by the renowned spiritual healer Harry Edwards and was picked from the audience to go on stage and assist. His interest in healing deepened and he continued to develop his abilities until, when he was 21, his spirit guide 'Paul' came through and they have been working together ever since. Just as with my own trance mediumship involving White Feather, Ray too is completely taken over by the guide and is not aware of anything that happens whilst Paul is in control. Complete and utter trust and love is essential for this type of work to take place, as is a complimentary level of spiritual development. If there did not exist a true spiritual affinity and agreement between guide and medium, then no 'point of contact' would be possible. I'm certain that Ray will agree with me when I say that even though the non-physical entity acting as the 'control' in trance mediumship (a very pure form of mediumship by the way) is more spiritually advanced than the medium, he or she is never so far ahead as to be unable to operate through their earthly instrument. As with the aforementioned psychic surgeons it seems that when entranced, Ray's guide knows exactly what the patients problem is, it's origin and the best form of treatment.

There are many excellent testimonials on Ray's website that verify the incredible work that he does. Here are just a couple that stand out:

Barbara Allen, the Mother of 21-year-old Steven Allen, contacted the Psychic News Newspaper on the 3rd July 2004 to confirm the miracle cure to Steven. Her words were as follows:

"My Son Steven was born with no food gullet and after a lot of suffering and 29 operations, surgeons managed to reconstruct an oesophagus and move his stomach into his chest area. The surgery included eight operations at London's world famous Great Ormond Street hospital for children. Until recently Steven was having a dreadful time swallowing his food, getting thinner and thinner, and having to physically press his neck to push his food down. He was terrified of going into hospital again to have more surgery. We visited his surgeon who suggested another barium meal X-ray, but the appointment was not for three weeks.

In desperation I phoned Ray's secretary and begged her to squeeze Steven in and get an appointment at his Leicestershire clinic. Earlier in the year I had seen Ray Brown in a Demonstration of Spiritual Surgery at a Seminar in Scarborough and I just knew that Paul - Ray's spirit guide -

would be able to help.

Paul performed an operation which he allowed us to video on Steven's reconstructed oesophagus and realigned his throat cage. Steven felt everything as Paul worked and had immediate relief.

Steven is now completely cured of the pain and eating normally. He can happily eat massive steaks, chips and doughnuts which he could never eat before.

Steven attended the hospital to have the barium meal test three weeks later, not telling the radiographer where he had been.

The X-rays confirmed that Steven's throat was normal when swallowing the barium meal. The radiographer started to wonder what was going wrong with his machinery and was very interested in our story when we told him about the spiritual operation.

We kept the appointment with Steven's surgeon and told him all about it. He was amazed. We are eternally grateful to Ray/Paul".

Rita Woodland was another who visited Brown and benefitted from the powerful healing force transmitted through him. She writes:

"I came to see Ray and Paul in August 2013. I had suffered a small stroke in late December 2012 and had been left with daily episodes of headache, numbness down my left side and visual disturbances caused by scarring to the brain (shown to me on an MRI at the hospital). I had been suffering daily with this for 8 months, even on the morning before my appointment. Paul said he would get rid of the scarring on the brain which was causing it and from that day I have not suffered anymore symptoms, they have completely gone. No more numbness, visual disturbances or headaches. Kind

regards and God bless you all."

A further glowing testimonial that proves the effectiveness of the healer's intervention comes from Susan Sharpe, who states:

"Thank you so much for the help you have given me since I have been coming to the Bury St Edmunds clinic. I first attended the clinic with severe endometriosis, which is a very painful gynaecological condition and is difficult to treat with orthodox medicine. When I first saw Paul, I was experiencing severe pelvic pain for almost the whole month, only having a few days relief before the pain started again. I had pain during the day, and would also wake in the night desperate for painkillers and a hot water bottle. Lack of sleep was also taking its toll.

I had been given an MRI scan at hospital, which showed an extensive and high level of endometriosis, and I was advised that an operation to remove my ovaries, tubes etc. was the best option. I really didn't want this. I had seen Ray demonstrate his work some years ago, and really felt this was something I wanted to try. From the first session at the clinic my pain levels reduced hugely, and with each session I felt better and better.

I now volunteer as a healer at the Bury clinic, so I feel very lucky that I can see Paul on a regular basis if I need to, and also support his work. There is no doubt in my mind that I have been able to avoid what I consider to be a major operation, and that I have got my life back. When I do see him now he keeps the endometriosis under control, and also helps me with any other aches and pains I have at the time.

I feel very blessed."

There is no disputing the sincerity of these and many other testimonials on Ray Brown's website, all of which give credence to his work and the genuineness of his ability and I know from some who have been to see him, that he is a credible healing channel.

The aforementioned Harry Edwards (1896-1973) was another remarkable exponent of the light, a gifted healing channel who helped countless thousands of humans and animals through both 'contact' healing and 'absent healing'. Although Edwards doesn't fit into the bracket of 'psychic surgeon' there is no doubt that his 'spirit helpers' performed many operations from a frequency beyond this physical one that we are so familiar with.

Classed as a 'spiritual healer' Edwards is of particular interest to me because in an indirect way he was instrumental in my own early spiritual development. My grandmother, who was bedridden for many years, regularly wrote to him requesting distant healing for herself and others whom she knew were in need of help. It was her discovery of spiritual healing and subsequent long-standing interest that lead to circumstances involving another healer from Canada named Colin Turner coming to see her whilst visiting the UK. Through this association I too received hands-on healing resulting in the opening up of my own spirituality. Indeed, so powerful was the healing experience from my one and only meeting with Turner, I can say without doubt that it was life-changing.

Edwards though, was involved in changing the lives of so many with his public demonstrations, private healing sessions and the setting up of the world famous Harry Edwards Spiritual Healing Sanctuary at Burrows Lea in Shere, UK. At the height of his powers the sanctuary was receiving 10,000 letters a week asking for help and is still operating effectively today, as I know from personal experience.

Edwards of course, was also a medium and thus acted as a conduit for the power and light that flowed through him, transforming those who were sick and often restoring them to full health. I have been one of those recipients as have other members of my family and I can say beyond reasonable doubt that whenever I have received spiritual healing (as it known) it has been accompanied by a sense of mental wellbeing and upliftment.

The crucial issue in all of these examples, which I cannot emphasise enough, is that whether or not mediumship is the central feature connecting those who are able to exhibit such incredible psychic and healing powers, they have all without exception evolved to a sufficiently high spiritual vibration to enable the divine light of all life to operate, aided or otherwise, through them. Whether they are vehicles for discarnate healers or whether their own inner light is the catalyst for change at the physical level, does not detract from this point. In fact, history will show that some of the most famous 'physical mediums' of the past were quite 'earthy' characters displaying obvious human traits and weaknesses. Yet at some higher 'soul' level, they must have been sufficiently advanced to enable the very constitution of their being to accommodate the higher energies necessary to facilitate the phenomena. Possessing an earthly personality is one thing, but the qualities of the infinite self are quite another. When we use terms like 'expanded consciousness', 'self aware', 'self-realised' 'awakened' or 'enlightened' it is important to make this crucial distinction:

It isn't that our mind or consciousness has become that [insert choice of term - eg: enlightened]
Rather, it is that the delusion that once concealed our understanding is no longer present.

If you've ever scraped the ice off your car windscreen or wiped your hand across misted glass, you'll know exactly what I mean when I say that you can see clearly what was before, obscured.

Mind over matter

History is littered with stories of mystics and gurus capable of amazing feats that defy logical explanation by mainstream science yet when seen from the perspective of infinite consciousness the picture becomes clearer. As we have already seen, India is a country enraptured with spiritual beings, where stories of miracles seem almost ordinary. One such proclaimed prophet, eighty-two year-old Prahlad Jani, who lives in the state of Gujarata claims he hasn't consumed any food or water in more than seventy years. Medical doctors have monitored Jani on more than one occasion and have gone on record as saying that during the time they were with him, he neither consumed food or drink nor passed any stools or urine. When it is considered that a human being can generally only survive for three or four days without water and a month without food it is clear that something seemingly incredible is going on and Jani has stated that his 'powers' come from God.

Although on occasions we can never entirely rule out trickery and as in all walks of life there are to be found charlatans and deceivers it is much more difficult to explain through conventional science how it is possible that Yogi Rambhauswami from the Indian village of Tanjore, was able to lie in a blazing fire, seemingly for hours at a time. Even though his beard was singed and his clothes smouldered a video taken of the event appeared to show the mystic apparently calm and serene amidst the tremendous heat that surrounded him as he lay cloaked in his saffron blanket. A monk explained:

"When you are with God, you will get the power. When you are not with God, you can't do anything."

When a person is sufficiently spiritually evolved to the point at which their own inner light and power functions without undue effort and conscious attention, what seems for most impossible, becomes normal. If we consider that the collective 'reality' that the majority of us are decoding has built into it a kind of 'freewill firewall' that prevents us from undertaking certain behaviours and actions, it seems logical to assume that what we perceive in a certain way will always appear as such, unless it is proven otherwise. For example, in our collective experience flame is always hot and will burn pretty well everything it comes into contact with. Walls are generally solid and impervious - we cannot pass through them, which is why we need doors. Water on the other hand, is permeable to the point where we cannot walk on it without sinking. All of these things are a given, as are the 'laws' of energy and matter with which they interact.

But, from a heightened perspective or a dimension outside of the electromagnetic spectrum in which we reside, the laws may be different or even non-existent. Walls may be transparent, water may support us and fire may contain no heat. If a mystic, swami, guru or indeed any one of us capable of being transformed by the light of infinite consciousness can hold a state, temporary or permanent in which the decoding mechanism *(body-mind)* can transcend the known 'laws' of physics, then the 'miraculous' can occur. Normality as we know it, along with all that accompanies it, can be circumvented by the emancipated being and as a wise soul once said 'both time and space can be obliterated by a single thought'. This is so true, but I would also add that both time and space as we know them are part of the illusion that the mind presents to us. Infinite consciousness has no such limitations imposed upon it and when fully realised in this blissful state of awareness, the impossible becomes possible.

Transcending Maya

According to some schools of Hinduism the world is an illusion, a play of the supreme consciousness of God. It is a projection of things that are impermanent, but give the impression of being permanent. In later Vedic texts and modern literature dedicated to Indian traditions, *maya* connotes a 'magic show, an illusion where things appear to be present but are not what they seem'. But this is somewhat misleading because maya is not just referring to delusion. The Vedic scriptures state that the material world functions according to the one fundamental law of maya - the principle of relativity and duality. Nature itself is maya, as are all aspects of the phenomenal world, where opposites appear everywhere we look – negative and positive, repulsion and attraction, heat and cold, darkness and light. It is easy to see why some interpret these states as illusions. When experienced from one level they are because at the frequency of infinite consciousness *there is no duality - all is one*. But here on Earth, where 'reality' is what we decode it to be through the *bodymind* 'computer', opposites exist, as do states of delusion and illusion. This dichotomy though, is resolved within the consciousness of one who is considered a master and the play of his inner light transforms matter into states of malleability denied to the many. Paramahansa Yogananda understood this principle and his writings attempt to convey the means by which spiritually evolved individuals create the miraculous. He commented:

> *'In his famous equation outlining the equivalence of mass and energy, Einstein proved that the energy in any particle of matter is equal to its mass or weight multiplied by the square of the velocity of light. The release of the atomic energies is brought about through the annihilation of the material particles. The "death" of matter has been the "birth" of an Atomic Age.*
>
> *Light-velocity is a mathematical standard or constant not because*

there is an absolute value in 186,000 miles a second, but because no material body, whose mass increases with its velocity, can ever attain the velocity of light. Stated another way: only a material body whose mass is infinite could equal the velocity of light.

This conception brings us to the law of miracles.

The masters who are able to materialize and dematerialize their bodies or any other object, and to move with the velocity of light, and to utilize the creative light-rays in bringing into instant visibility any physical manifestation, have fulfilled the necessary Einsteinian condition: their mass is infinite.

The consciousness of a perfected yogi is effortlessly identified, not with a narrow body, but with the universal structure. Gravitation, whether the "force" of Newton or the Einsteinian "manifestation of inertia," is powerless to compel a master to exhibit the property of "weight" which is the distinguishing gravitational condition of all material objects. He who knows himself as the omnipresent Spirit is subject no longer to the rigidities of a body in time and space. Their imprisoning "rings-pass-not" have yielded to the solvent: "I am He."

"Fiat lux! And there was light." God's first command to His ordered creation (Genesis 1:3) brought into being the only atomic reality: light. On the beams of this immaterial medium occur all divine manifestations. Devotees of every age testify to the appearance of God as flame and light. "The King of kings, and Lord of lords; who only hath immortality, dwelling in the light which no man can approach unto."

A yogi who through perfect meditation has merged his consciousness with the Creator perceives the cosmical essence as light; to him there is no difference between the light rays composing water and the light rays composing land. Free from matter-consciousness, free from the three dimensions of space *and the fourth dimension of time, a master transfers his body of light with equal ease over the light rays of Earth, water, fire, or air. Long concentration on the liberating spiritual eye has enabled the yogi to destroy all delusions concerning matter and its gravitational weight; thenceforth he sees the universe as an essentially undifferentiated mass of light."* The sage continues:

"Optical images," Dr. L. T. Troland of Harvard tells us, "are built up on the same principle as the ordinary 'half-tone' engravings; that is, they are made up of minute dottings or stripplings far too small to be detected by the eye. The sensitiveness of the retina is so great that a visual sensation can be produced by relatively few Quanta of the right kind of light." Through a master's divine knowledge of light phenomena, he can instantly project into perceptible manifestation the ubiquitous light atoms. The actual form of the projection—whether it be a tree, a medicine, a human body - is in conformance with a yogi's powers of will and of visualization.

In man's dream-consciousness, where he has loosened in sleep his clutch on the egoistic limitations that daily hem him round, the omnipotence of his mind has a nightly demonstration. Lo! there in the dream stand the long-dead friends, the remotest continents, the resurrected scenes of his childhood. With that free and unconditioned consciousness, known to all men in the phenomena of dreams, the God-tuned master has forged a never-severed link. Innocent of all personal motives, and employing the creative will bestowed on him by the Creator, a yogi rearranges the light atoms of the universe to satisfy any sincere prayer of a devotee. For this purpose were man and creation made: that he should rise up as master of maya, knowing his dominion over the cosmos.'

Knowing what we do about the nature of light, let's analyse

this information. What does Paramahansa mean when he states *'only a material body whose mass is infinite could equal the velocity of light'*? According to conventional science, the speed of light cannot be achieved because as an object (or human) accelerates, its mass increases. In science, mass is just simply 'locked up' energy. Another name for this type of mass is 'inertial mass.' Essentially, inertial mass is the amount of resistance that a physical object has to any change in its motion (this includes the resistance that a body has to acceleration or to directional changes). According to Einstein's theory of relativity, gravitational mass is always equal to inertial mass. But when we speak of an object's mass increasing due to acceleration, we are really talking about an increase in its inertial mass. So, if we think of mass as energy, we can see why an object will experience an increase in mass as it gathers speed along with the amount of energy it has. This understanding of mass and energy makes it easier to understand one of the reasons why (according to mainstream science) we can never reach the speed of light. The energy of a particle diverges to infinity as it approaches the speed of light and since a particle can never have infinite energy, the speed of light cannot be attained. Since an object has infinite kinetic energy when it approaches the speed of light, it therefore has infinite mass as well.

But Paramahansa gives us another clue when he writes *'The consciousness of a perfected yogi is effortlessly identified, not with a narrow body, but with the universal structure.'* Whilst the majority of people think in terms of three or four dimensions and consider physical matter to be solid, existing in time and space, the self-realised yogi encounters no such constraints. When he refers to 'the universal structure' Paramahansa does not imply physicality or the associated forces to which it is subjected, but rather what I and others refer to as the 'meta-physical' universe, comprising of quantum fields and waveforms that are in essence *infinite possibility* awaiting the interaction of consciousness to collapse into 'reality'. It is the light of consciousness that chooses how the formless will become the form. It is the light of consciousness, operating through mind that decodes the waveform *information* into the particle. To me, it makes perfect sense that one whose inner light and consciousness has evolved and expanded to the point of self-realisation should also be able transcend the boundaries and constraints of the material universe, attaining mastery over creation. In fact I would go further and state that this divine gift is an integral part of true spiritual unfoldment. Dissecting the wise words of the yogi further it is clear that light is the key element employed in the manifestation of form; *'Through a master's divine knowledge of light phenomena, he can instantly project into perceptible manifestation the ubiquitous light atoms'* (Fig 26.)

Gifts of the Light

As the study of quantum mechanics has already proven that 'reality' doesn't exist until consciousness collapses the waveform information into particle form, we are already participants to a point, in what we experience. So it doesn't take much of a leap of faith to conceive of a consciousness so spiritually developed and a mind so powerful that it is able to project its own inner light energy in such a way as to be able to manifest at will, whatever it desires. That the majority of people as yet cannot perform this feat does not disprove its validity, but rather, is a reflection of their lack of true spiritual development at this level of being.

The comment that *'Gravitation, whether the "force" of Newton or the Einsteinian "manifestation of inertia," is powerless to compel a master to exhibit the property of "weight" which is the distinguishing gravitational condition of all material objects. He who knows himself as the omnipresent Spirit is subject no longer to the rigidities of a body in time and space.'* is suggestive of the self-realised master being above and beyond certain 'natural laws' operating throughout the physical universe and of course, this makes perfect sense when there are no 'physical laws' at the level of infinite consciousness and all-possibility. When the level of expanded consciousness has been attained, one is connecting deeply with infinite potential, where all conditioned thought and belief are transcended. Here, *all* is possible and as the mind projects at will its desire to create, so then does it manifest. But again, I must stress that although this is possible for all, it is not so for 99.99999 per cent of us *at this point in our evolution*. Only a relative few within this frequency range can manifest this level of energy and light, not because they are greater or better than the rest of us, but because they have attained this level of expression on their 'journey'.

If at this point, you are starting to feel excited about your own possibilities, wondering how you can start to move towards the light and open yourself to the infinite potential that lies within, there is good news and bad news. The good news is that all attempts to improve spiritually are met with proverbial open arms by infinite consciousness and every effort that is made, as long as the intent is pure and selfless, is reflected in a quickening vibration and expanded state of being. The bad news is that there are no quick fixes or short cuts. You can't download the manual or buy an instruction book - there isn't one. This is why the power behind real manifestation comes with conditions attached that I like to think of as a kind of safety mechanism, a bit like brakes that trip in on a runaway train, preventing it from going off the rails. Artificial 'power' requires artificial support - technology, people, and money and the right circumstances. *Real* power simply requires just you and your true, selfless desire to serve others without expectation of gain. The greatest gifts in life - health, friendship and love, cannot be contained. They exist naturally but need to be afforded respect and attention to enable them to flow freely. For love to be reciprocated it has to be sent forth. So too friendship, whilst good health is generally the result of considered living – balanced diet, exercise and moderation in all things. Similarly, the emergence of the inner light is a gift that can be encouraged but cannot be forced. It increases when conditions are suitable, but ultimately has to expand beyond anything that contains it, gravitating even beyond the nature of the individual, in much the same way as each of us are connected to and exist within a particular resonant field.

For its purpose to be fulfilled, light must illuminate the way and radiate outward to all within its orbit acting as the catalyst for greater human spiritual evolution. From a personal perspective, in creative terms the upshot of this is that just as certain mental and physical attributes develop through the inevitable consequence of 'physical' growth and as wisdom emerges from years (or even multiple lifetimes) of experience, so control over the very 'substance of being' is the ultimate consequence of true spiritual expansion. This is how spiritual masters are able to walk through 'solid' walls, manifest fragrances, create 'forms' from thin air, walk on water and heal the sick.

It is only natural that when we glimpse sight of these kinds of possibilities, we want them for ourselves. Just as an infant is attracted by a shiny toy, twirling round on a string suspended over its head or captivated by a kitten chasing a ball, so too do those of us who, having passed through the infancy stage of the school of spiritual development, desire things of more substance and meaning.

That which once held our interest and entertained us psychically will have inevitably given way to thoughts of something deeper and the need to understand the real purpose of life. Messages from mediums, describing loved ones who have passed, however entertaining, evidential or comforting, will have satisfied us only briefly, leaving us with the desire to know still more and perhaps even develop our own communicative abilities, that we can see or speak with them directly. Our increasing sensitivity, painful at times, may have left us longing for another life, away from the darkness and despair of this world. We may have grown weary of our lifestyle and our place of work and maybe even some of our friends with whom we no longer seem to connect in the way that we once did. This is entirely natural and we should not resist it.

And so our thoughts and aspirations take us beyond ourselves, towards the intangible, ethereal, yet increasingly more attractive non-physical planes that we sense are all around, waiting in readiness for us. Energies stir within us and we sense the start of something special, gaining momentum and directing us, carrying us forward into a new arena of thinking. We are drawn to philosophy and to books containing answers to the many questions that constantly arise. Then amidst this restless wonder and need to know, a deep peace begins to emerge, along with the deeper yearning to serve others and share with them what we know and have yet to discover. This truly is *the dawning of enlightenment.*

6

Another Perspective

"If you have a well in your backyard that you get your water from and I have a well in my backyard that I get my water from, we both are going to go to separate wells to get our water, but if we know anything about wells, we know that the deeper we go the source of water is the same and that's the unifying force."

Saul Williams (1972-)

After I had completed the manuscript for *Transcognitive Spirituality* and the book was about to be published, I became aware of some research that had been done suggesting that the universe in which we live might actually be a 'computer simulation'. I didn't have time to include the information in the book beyond just a short postscript alluding to the fact I was aware that the concept existed and believed that it was worth serious consideration. Since then I have looked into this remarkable claim more deeply as additional information has emerged from several sources.

I knew that when I began writing this book, I would at some point have to address this issue and discuss the facts surrounding it.

As this work is primarily about light it seemed to me probable that this would feature somewhere in any theory describing the nature of the cosmos, at least in terms relating to its physical laws and structure. But it is highly likely that any kind of artificial simulation, were it to exist, would also impact upon the inner light and our human spiritual connection to higher frequencies, because everything is connected.

Added to this, the whole concept of a simulated universe leads to another point that I feel I have to address - one that challenges some of the fundamental issues surrounding the role of visible light in both our limited 'physical' reality, as it relates to our eternal existence - especially in regard to the 'restrictions' on human perception imposed by it. But firstly, let's look at the information emerging in relation to the so-called simulation.

There is an old Chinese parable that speaks of an emperor dreaming that he was a butterfly, dreaming that he was an emperor. In modern terms, it might be 2070 with people living in a computer simulation of what life was like in the early twenty-first century. Alternatively it might be millions of years from now, and people are in a simulation of what primitive planets and people were once like in 2016 or at any other period in linear time, a bit like the 'holodeck' featured in *'Star Trek; The Next Generation'*. So how can we be entirely sure when and where we really are? The probable answer at this point is that we cannot.

Swedish philosopher Nick Bostrom (1973-) from Oxford University, UK describes a fake universe as a *"richly detailed software simulation of people, including their historical predecessors, by a very technologically advanced civilization."* He doesn't state that we *are* living in such a simulation, only that it is feasible that we *could* be. *(Fig 27)*.

Fig 27.

His philosophical point is that *all* cosmic civilizations either disappear (destroy themselves) before becoming technologically capable, or *all* decide not to create whole-world simulations (decide such creations are not ethical, or get bored with them). But the key word is 'all', because if even one civilization anywhere in the universe were to generate such a simulation, then simulated worlds would multiply rapidly, meaning that almost certainly humanity would exist in one. This theory is based on the fact that computing power grows at an exponential rate, a point not missed by some visionaries and scientists who think that if we are within some kind of computer generated simulation, it would be so authentic that we would be unlikely to know that it existed. This would also mean that everything is software, right down to every single atom in our bodies and maybe even our consciousness. Perhaps the only way that we might know is if the programmer made a few errors or even left some clues, deliberately or otherwise, enabling us to find out for sure. Interestingly a recent un-peer-reviewed attempt at a more rigorous physical analysis of the situation, *Constraints on the Universe as a Numerical Simulation* by physicists Silas R. Beane, Zohreh Davoudi, and Martin J. Savage, reached the conclusion that *"in principle there always remains the possibility for the simulated to discover the simulators."* Their specific prediction was that there might be limitations on cosmic ray energy levels if we live in a simulation.

Strangely enough, there are some clues, but the question is 'how reliable are they?' As we know it can be easy to misinterpret readings or make the information fit into a specific model (global warming comes to mind) and we have to be cautious before jumping to the wrong conclusions. Another scientist, physicist Paul Davies (1946 -) draws upon 'simulation theory' to suggest that:

"If you take seriously the theory of all possible universes, including all possible variations, at least some of them must have intelligent civilizations

with enough computing power to simulate entire fake worlds. Simulated universes are much cheaper to make than the real thing, and so the number of fake universes would proliferate and vastly outnumber the real ones. And assuming we're just typical observers, then we're overwhelmingly likely to find ourselves in a fake universe, not a real one."

Theories are one thing, but facts are quite another of course and science prefers hard evidence to support theoretical concepts. To this end, Beane developed a 'cosmic ray test'. This involved scientists building up a simulation of space using a lattice or grid *(Fig 28)* having calculated that the energy of particles within the simulation is related to the distance between the points of the lattice and that the smaller the lattice size, the greater the energy that the particles can have. A paper published in November 2012 calculated that the 'arbitrary cliff' in the spectrum, consistent with the kind of boundary that would exist if there truly were an underlying 'lattice' governing the limits of the simulation had indeed been found. In simple layman terms this is like the analogy of a ball being thrown an enormous distance before eventually falling over the edge of a cliff - in this case the extremity of the simulation or matrix.

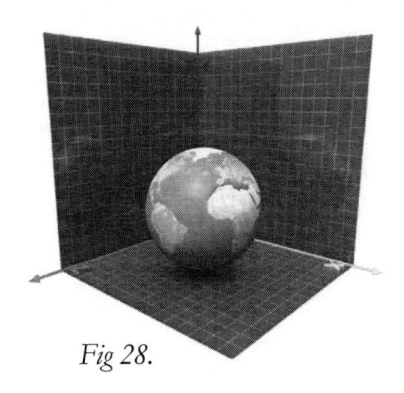

Fig 28.

There is also the possibility that the universe is pixilated. Get close enough to your computer or TV screen and you'll see pixels, small points of data that make for a seamless image when viewed from a distance. Scientists think that universal information may be contained in the same way, and that the natural 'pixel size' of space is roughly ten trillion trillion times smaller than an atom, a distance that physicists refer to as the Planck scale. However just when it seemed that the rule book would need to be rewritten, scientists from the Fermilab Holometer, located in Illinois, USA, announced that they had 'ruled out' the theory of a pixilated universe to a 'high level of statistical significance.' I'm not sure about this though and I don't think we should be too quick to accept it without further proof. In any event it doesn't prove that the universe isn't holographic - because it is.

We have to be mindful too, that most of the research carried out by scientists, whilst supposedly independent, is seldom so, with funding almost always provided by private or governmental sources. Far be it from me to suggest that the 'system' that has invested so much in maintaining the 'this world is all there is and we are alone in the universe' paradigm has more to gain from keeping us in ignorance than from allowing the truth to be revealed, but I will anyway.

Another freethinker, writer and researcher David Icke (1952-) has a slightly different take on it, which brings me to the second point of discussion. For many years he has suggested that we live in an artificial simulation as a result of the waveform construct having been 'hacked' by a race existing within another frequency range very close to our own (for purposes that we won't go into right now) and his contention is that visible light forms the boundary of the energetic 'prison' in which we exist. He suggests that it is the periphery of our 'artificial reality' and that because we cannot seemingly travel faster than the speed of light, we cannot escape it. Even more worryingly is his assertion that light forms part of the illusion that traps us in the afterlife, resulting in multiple incarnations (reincarnation) on Earth over which we have no control. In his own words:

'It's my view that if you 'go to the Light' [after death] you will stay in the Matrix to be recycled back to five-sense Earth for another lifetime of slavery, no, oops, sorry, another incarnation on your endless journey to learning on the road to enlightenment.'

It's a thought; I'll grant him that, but exactly how plausible I'm not certain.

If you are still with me at this point, keep an open mind because I know that this is going further down the rabbit hole than you may have anticipated (which isn't a bad thing) but with good reason. Continuing the Icke theme for a moment, in a recent book *Phantom Self* he writes:

"There is widespread confusion between light (as in the light that allows you to see in the dark, sunlight, lamplight) and the frequency band of visible light and the electromagnetic spectrum. Scientists talk about dark matter and dark energy, but these are terms that only mean they are invisible to us (of another frequency) and not that they are pitch black – hey, Ethel, got a match? When you are resonating to the same frequency as your reality you

can 'see' just as well as you can in visible light. It is not about 'light' in terms of luminosity, but the frequency band of what we perceive as light – the frequency band, or one of them, of the Matrix [simulation]. The light is not the way out of here - it's the trap."

Now that again is pretty strong stuff but I understand where he's coming from and in some ways he's exactly right - we are relatively 'blind' in terms of only being able to register the small range of electromagnetic radiation that is visible light. Also, many things, even those that appear positive and good, can trap us. To paraphrase Freddy Mercury, 'Too much love will kill you'. In most cases it won't of course and neither will light, but I can see how either could be considered as a trap. Water is essential to life, but you can drown in it. Food helps regenerate the cells, but it can be poisonous. We need air to breathe, but it carries bacteria and viruses that can kill – you get the idea. It is true that certain aspects of nature can and do act as energetic barriers, probably for good reason. If we were open to every radio frequency for instance, without having an in-built mechanism that tunes out the sound - which is why we need a radio tuner to convert the electromagnetic waves back into sounds that we can hear - life would be unbearable (in fact, come to think of it listening to some of the current radio stations is also unbearable, but that's just me). So does this make 'sound' a prison wall to humans? Just because we are unable to hear what a dog can, does that imply that we are unfairly handicapped? Maybe, but I don't see it that way.

I do respect Icke's views though and what he has accomplished in both pushing forward the boundaries of understanding relative to the metaphysical universe and at the same time exposing the agenda of human manipulation and control that confines our understanding of the true nature of life. The upshot is that whilst I'm open to some of his ideas about light, I'm yet to be convinced by all of them, particularly when it comes to the 'afterlife'.

Although Icke suggests we exist within a 'fake reality' in which light acts as a firewall, keeping us in ignorance of the true infinite awareness that we really are, importantly he also maintains that we remain connected to our higher self *because we are our higher self.* On this point, I totally agree. Whether or not we exist in an endless 'time loop' that sees us reincarnate over and over again into a false reality in which everything is deliberately inverted in order to keep us trapped, we are still *infinite consciousness* and as such are eternal expressions of this, albeit in degree. Awareness in the purest sense is what it says on the tin – *pure awareness.* It is only when that awareness experiences itself through a human body that its essence appears diminished. As I have said, it is the filter of the mind and all that accompanies it – the sense of 'self', of individuality, of linear time, of identity, of detachment (to higher states), of attachment (to lower states), of personality, of ego and everything else that detracts from the pure expression of all that is. Whether or not the reality we experience is *illusion,* whether or not we do live in a simulation, whether or not the universe is holographic, *we are ever connected to all that is, ever was and ever shall be.* The problem, if I can refer to it as such, is that the denseness and conditioning imposed by life at the more fundamental levels is so gross, we forget what we are. More than this, we actually believe that we are something and someone that we are not.

Approaching transcendence

Relative to the inner light, which is not dependent on the electromagnetic spectrum or whether or not we live in a simulation, there are several determining factors that can hold back its expression, which I will come to shortly. Recognising these blocks is an important part of self-development and is worth more than anything in terms of freeing oneself and allowing movement towards true enlightenment. To reiterate, it's not a case of simply 'asking' the universe or 'believing' that what we desire will arrive, we must have achieved a level of true spirituality - and by that I don't mean the attainment of certificates, qualifications or any religious status – that enables the light and its accompanying power to manifest through us in ever greater degrees.

As we saw from David R. Hawkins' Map of Consciousness, higher levels of awareness and enlightenment are concomitant with increased inner light and energy curves and by adopting certain lifestyle changes allied to specific regimes designed to 'quicken' the individual frequency at which we resonate, sooner or later new vistas open to us. More popular disciplines, outside of any religious doctrines, include mindfulness (very popular at the moment), meditation, contemplation, yoga (various kinds), sitting for the development of mediumship (my initial route) and retreats (especially silent ones). Other techniques involve focusing upon the 'body' as in fasting, dieting, exercise and breath work, all of which are tools that can aid us to a point. Some of these approaches are more beneficial than others, depending upon both the physiology and mental state of the individual and ideally, a combination of disciplines, provided they do not conflict with each other, can be beneficial. Whether these approaches are undertaken alone or with the support of a group is down to each individual but the intellectual approach alone will never reap the desired outcome. Always, positive change must emerge naturally from within and *spirituality has to develop outwardly from within, never inwardly from without.* Reading books, going to lectures, attending workshops and gaining qualifications might look good on the CV, but it is what develops internally, the inner light, that makes for true enlightenment.

Many years ago I predicted the coming together of spirituality and science and just as in my last book, this one is a fusion of both ancient and modern, scientific and esoteric approaches. One such exponent of this approach that I find highly credible is John Hagelin (1954 -). Hagelin *(Fig.29)* is an American particle physicist, author, lecturer, three-time candidate of the Natural Law Party for USA President and also director of the Transcendental Meditation movement for the USA. He is also Professor of Physics and Director of the Doctoral Programme in Physics at Maharishi International University. After completing his Ph.D. Harvard University in 1981

Fig.29

Dr. Hagelin joined the theoretical physics groups at the European Laboratory for Particle Physics (better known as CERN) and the Stanford Linear Accelerator Center (SLAC), where he was actively engaged in fundamental research at fundamental research the forefront of 'supersymmetric' Unified Field theories. Following his positions at CERN and at SLAC, he joined the faculty of Maharishi International University, where he established a doctoral program in elementary particle physics and unified quantum field theories. He has published extensively in the area of supersymmetric unified quantum field theories in such journals as *Physics Letters, Nuclear Physics* and *The Physical Review.*

In 2015 I watched a lecture that he presented for Stanford University and was delighted at the way in which he incorporated cutting edge scientific knowledge and ancient spiritual wisdom. In his view consciousness can be thought of as part of the Unified Field of all life, lying deep within the self, beyond thought. *(Fig 30).*

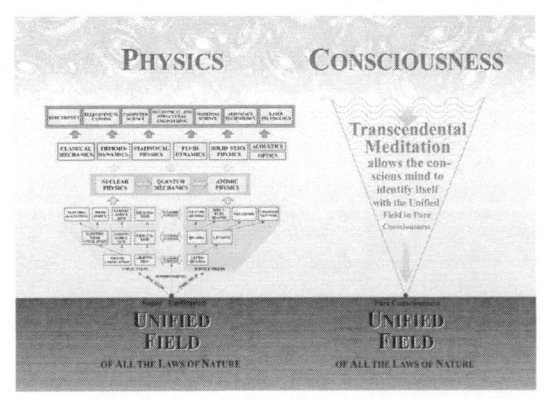

Fig 30. John Hagelin's schematic - Physics, and the Unified Field

I couldn't agree more and would emphasise that it also exists and extends *beyond* the self, but that's just semantics. What really grabbed my attention is his suggestion that consciousness is 'structured in layers' parallel to the 'physical' universe. He states that when we access deeper, more expansive levels of thought or even go beyond thought, we are connecting directly with the Unified Field. Again, I agree.

Hagelin went on to explain how in scientific terms nature exists in layers from *'fundamentally unified to superficially diversified'*. These four layers are:

- **Classical** – our everyday world and experience
- **Quantum Mechanics** – atoms and molecules
- **Quantum field Theory** – sub atomic particles and forces
- **Unified Field Theory** – the underlying field that contains a type of programming code for all the structures of creation

If you are at all familiar with *Superstring Theory* (the attempt to explain all of the particles and forces of nature in one theory) you might already know that superstrings are *vibrational energies* (that we used to call particles) that emerge from the Unified Field *(Fig 31)*. They are not strings as such but have the characteristics of rubber bands and are infinitesimally diverse in size.

In accordance with their particular energy, they correlate directly to different types of particle and effect, ie: gravity, light, etc. Therefore the emergence of superstrings from the Unified Field directly creates the universe and its laws that we observe and experience. To put it another way, the Unified Field is *infinite consciousness* and superstrings are an effect of consciousness that can be seen and measured. They are one way that coded information is expressed between higher and lower frequencies throughout the universe. This knowledge gives fresh insight into exactly how mystics, swamis and other enlightened beings manifest their 'miracles'. It is *their* expanded consciousness (inner light) interacting with infinite consciousness (all possibility) to manipulate energy from the waveform level into the holographic reality that we experience as form. The catalyst in the information transfer between the consciousness of the yogi and the universal consciousness is - *light*.

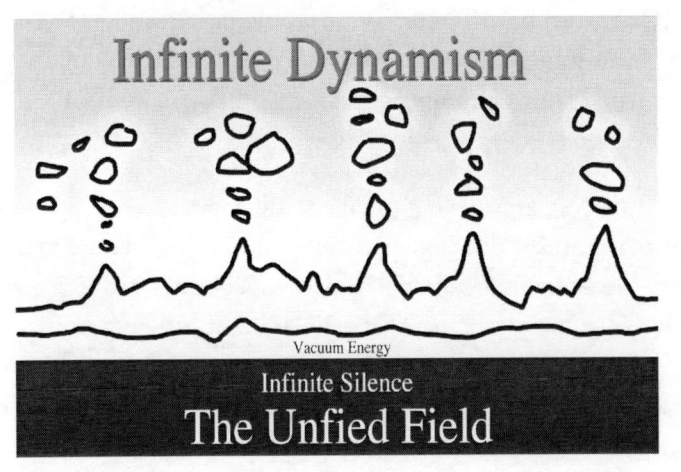

Fig 31. John Hagelin's schematic - Superstrings emerge from the Unified Field

In the lecture I saw, Hagelin also compared the deeper layers of the mind with mathematics, because both have layers of greater subtlety, concreteness and power. I suggest by the way, that mathematics was not invented by man, but rather discovered as being a fundamental part of the universal code, embedded throughout all that exists. If, when applying mathematics to scientific theories we add fractions, integers and so forth, these open up greater possibilities to understanding the nature of reality and the same can be said of the mind. The deeper one goes, the more peaceful and quiet that the mind becomes but significantly the power of creativity increases because one is more in touch with pure consciousness.

I've mentioned how I am guided to information and not long into writing this chapter I found a paper, written by Hagelin and published by Maharishi International University Fairfield, Iowa. In it he lays out in great detail his theories detailing the connections between consciousness and the Unified Field, but one section stood out to me when I read it, suggesting that my own conclusions about the inner light of consciousness and the manifestation of form were similar in many ways. Referring to the work of Maharishi Mahesh Yogi and the ancient science of consciousness known as Vedic science, Hagelin writes:

"The Vedic tradition also holds that it is possible for the individual to experience pure consciousness - the essential nature of consciousness itself. For this to occur, individual consciousness must be allowed to experience its "pure, self-interacting state," in which consciousness is awake only to itself, rather than identified with objects of perception, thought, or feeling. In this

state, the knower, the process of knowing, and the known are said to be unified, since consciousness itself constitutes both the knower and the content of experience (Maharishi Mahesh Yogi, 1985, pp. 64-66). A systematic refinement of the functioning of mind and body is said to be necessary for this experience to take place, and a set of procedures for such a refinement is described in the Vedic texts (Maharishi Mahesh Yogi, 1966, 1969; Patanjali, 1978)."

Note the words 'refinement of the functioning of mind and body' or as I would describe it; *the unfoldment of the inner light.* According to Vedic science there is no separation between the observer and the observed and any such appearance is due only because of a particular perspective of the gross sensory level (the way in which the mind decodes information into our holographic reality). According to Maharishi Yogi any such concept must be abandoned if a fully realised understanding of self and its environment is to be attained. The Unified Field (all possibility) is considered to be a level of reality at which any such separation cannot be inferred. As Hagelin states:

"The experience of the Unified Field of Consciousness in which the observer, the process of observation and the observed are unified, is considered to be a means of realising the ultimate inseparability of the observer and the observed, leading to a completely unified view of self and the environment traditionally known as 'enlightenment' or 'unity consciousness' (Maharishi Mahesh Yogi, 1966, 1985)."

Another Perspective

It is evident that Hagelin has familiarised himself with the works of Mahareshi Mahesh Yogi (1918 -2008) and is one of those rare individuals who can make the obvious connections between modern scientific understanding of quantum fields and ancient Vedic wisdom. Maharishi developed the technique of Transcendental Meditation (TM) as a means of allowing greater access to deeper levels of consciousness and Hagelin frequently refers to this during his lectures, acknowledging the power of this approach. Indeed, TM has been referred to as consistent with a *fourth major state of consciousness*, having aspects of both heightened awareness and deep rest, together. According to Mahareshi, it is when this pure, unified state of consciousness is attained that individuals can access directly the Unified Field and all of the laws of nature.

In what has come to be known as the 'Mahareshi effect' or *Super Radiance*, studies have shown that coherent energy produced by large number of people practicing TM can positively influence both the surrounding population and to an extent, events that occur (or don't occur, as in the case of violence) within that group. For instance, crime rates have been shown to decrease markedly with a drop of around one percent in cities that were targeted, as measured against non-targeted ones. This is unsurprising when you understand the power of the mind and the light energy produced by coherent thought. At one level, everything is energy, information and light and it makes perfect sense that with large groups of people connecting together in a common quantum field, the outcome has to be a positive, beneficial one.

Higher states of consciousness and Sidhis

Anyone considering the evidence for the connections between pure consciousness and the Unified Field will need to take into account what Vedic science calls *Sidhis* or 'supernormal abilities' that are associated with higher states of human development. Hagelin suggests that *'from a purely developmental standpoint, it seems reasonable to expect that new behavioral competencies would accompany higher developmental stages.'* That's scientific speak for what I've already outlined as an increase in the inner light and human energy curve that accompanies the expansion of consciousness to enlightened states.

A point worth mentioning here by the way, is that Paramahansa Yogananda stated that those producing materialisations and phenomena associated with Sidhis were not necessarily evolved spiritually. On the surface this would seem to contradict what I am suggesting in respect of these abilities only being attainable by those having reached higher levels of spiritual development. However, I would suggest that although one can be highly spiritually evolved, whilst expressing through the identity associated with the grosser form, this may not appear outwardly evident. Indeed, I have personally known some excellent exponents of mediumship whose personality traits left much to be desired, but these did not seem to prevent them from undertaking their spiritual work. The reason for this I suggest, is that the spiritual gifts that pertain to these individuals are the result of previous 'soul growth' and attainment rather than from any 'physical' constitution they might possess in their current incarnation and it is from this level or frequency that their true power originates.

Returning to Hagelin's paper, he outlines the work undertaken by developmental psychologists who have considered the possibility of continued growth to higher developmental stages 'beyond formal operations'. The four states of consciousness beyond the normal waking, dreaming and deep sleep described in Mahareshi Vedic Science are briefly summarised as:

1) **Pure consciousness** - the unified ground state of consciousness in which consciousness is identified with the Unified Field.

2) **Cosmic consciousness** - in which the experience of pure consciousness is permanently established along with waking, dreaming, and deep sleep states of consciousness. In this state, consciousness maintains its identification with the Unified Field whilst the mind and emotions are fully engaged in activity.

3) **Refined cosmic consciousness** - similar to cosmic consciousness except that the functioning of the mind and senses has become further refined. Sense objects are perceived in their most refined values and the emotions are said to achieve their full development.

4) **Unity consciousness** - in which the object, as well as the subject, is experienced as the Unified Field.

In this model the higher developmental stages (2-4) are said to develop spontaneously and are also said to form the basis for Sidhis that involve the ability to utilise the mind, body and environment in increasingly fundamental ways. The witnessing of some of these Sidhis, which appear to violate the known laws of classical Newtonian physics, would leave little doubt of the connection between pure consciousness and the Unified Field. The reason they 'appear' to violate the classical laws though, is because the 'laws' as we understand them are incomplete. They may have stood for centuries, but they are not comprehensive enough to deal with the latest understanding of 'reality' revealed through quantum mechanics. Newton may have been a genius in his day, but what did he know of the holographic nature of the universe or how the light of consciousness interacts with waveform information to collapse the wave into the particle that we experience as reality? Not a lot, I suspect.

So whilst phenomena such as levitation (with its implied control over the local curvature of space-time geometry) walking through apparently 'solid' walls and never consuming food might be off the scale for those coming from the understanding of classical physics, *when seen from the perspective of infinite consciousness and all that is, things appear quite differently.* From this expanded viewpoint we can recognise that things that before appeared miraculous, have their origin in the connection between individual (or collective) consciousness and infinite consciousness - *they are one and the same.*

Referring once more to the ancient saying 'as above, so below' we can see that light is light, whether it is from the Sun or from a battery torch. Ultra violet, infra-red, visible, non-visible, frozen, outer, inner - *it's all light.* And it's the same with consciousness, which is also light. Whether it is unity consciousness, fully enlightened consciousness, cosmic consciousness, refined cosmic consciousness, unity consciousness or even consciousness that is yet to awaken and become self aware - *it is one and the same.*

A flame is a flame, whether it is at the end of a match or leaping forth from the surface of a star. To our human acuity a match appears infinitesimally smaller than the Sun and so we accord the Sun with having much more 'power' than the match (at one level it *does*) but when seen from another perspective things may look different. For example, if our consciousness were to be contained within the perspective of a small insect held close to the match flame, it is likely that it would accord this far more power and importance than the Sun, apparently residing a great distance away in space - and who would blame it? Yet the potential within the flame, at the purest level, is equal to that of the Sun - it has to be - so too with consciousness. We are all conscious, we are all infinite consciousness, but in degrees of brightness relative to 'where we are'.

Film director, actor and author David Lynch (1946 -) famously said:

"If you have a golf-ball-sized consciousness, when you read a book, you'll have a golf-ball-sized understanding; when you look out a window, a golf-ball-sized awareness; when you wake up in the morning, a golf-ball-sized wakefulness; and as you go about your day, a golf-ball-sized inner happiness. But if you can expand that consciousness, make it grow, then when you read about that book, you'll have more understanding; when you look out, more awareness; when you wake up, more wakefulness; as you go about your day, more inner happiness."

Expanding our consciousness, or rather allowing an already expanded infinite consciousness to have self-expression, is also to facilitate the inner light and everything that accompanies it. We often say in jest, "the impossible I can do, but miracles take a little longer", yet when we attain the state of enlightenment the mystery of both are known to us.

Light and shade

I mentioned 'blocks' at the start of this chapter and there are some that can hold us back and to an extent temporarily diminish our inner light. Amongst these are fear, apathy, closed-mindedness, a distorted belief system, emotional or mental issues, physical pain and deeper spiritual/karmic issues. Fear is probably the most debilitating of these, manifesting through a number of emotional and mental states within the mind. Fear is the antithesis of harmony and peace, disturbing the balance of our inner being on so many levels.

We live in holographic reality that we believe is real. At this frequency level, fear also appears very real to many of us and believe it or not we can actually download fear, like a computer programme (or virus) along with other negative states, through exposure to things that instigate these changes within us. Occasionally these slip in quite subtly, under the radar of our conscious mind, through subliminal means but often they are obviously and blatantly clear such as when we are traumatised by violence on TV or in 'real life'. At times such as these, like a rabbit caught within the headlights of a car (now that is a light trap if ever there was one) we can become transfixed by the negativity generated by fear and literally freeze in the moment. How often have you found yourself unable to think clearly, speak or even move, through fear? It's happened to me a few times, but now that I know why, it doesn't have the same hold over me because I recognise that I have the power to change my perception of it.

Returning again to Hawkins' Map of Consciousness, fear comes pretty low on the scale and from a vibrational level there is an enormous gulf between fear and enlightenment. The truth is, fear and many of its close cousins - anxiety, depression, apathy, self loathing, anger, worry, despair and so on, all have one thing in common - they resonate at a low frequency and are not conducive to the expansion of the inner light. That said, they have no true power over light - only that which we afford them and *believe they have.* The truth is that they are all imposters of sorts - uninvited guests gate-crashing the party and even though they come bearing gifts (yes, they are teachers too) in the long term we wouldn't want to become best friends. The key is recognising them, which can sometimes be difficult, before they impact on us.

The 'world' that we experience in our head as being 'out there' is for the most part, anti-*light* and anti-*enlightenment* and it is manufactured by the few to be that way. From the lies and spin of politicians and state controlled media, planes flying overhead spewing chemtrails that block out the essential light of the Sun and do goodness knows what other harm, to man-made wars, genetically modified food, genetically modified people, 24/7 surveillance, human microchipping, corporate corruption on a global scale and the plan for a New World Order that serves only an 'elite' few (that combined, have more wealth than the rest of the world) it is clear that for most people, thoughts of survival are more paramount than aspiring to higher things.

As I have found to my cost, wise words are often wasted on people that walk in the shadows and have no desire to look up towards the light. The 'system' seduces the majority of minds through sport, gambling, drugs, alcohol, cigarettes, sex and entertainment whilst 'personalities' that serve the system are purposely promoted, revered and rewarded through perceived monetary gain, fame, material 'wealth' and enhanced social status. Conversely, those who question the system and know the real agenda and where it's planned to go are either ignored, ridiculed or both. In some cases, those exposing the global agenda are 'taken out' because they know too much and if left to their own devices would threaten its very existence.

If you have ever wondered why the world is as it is, why it's so difficult at times to even survive, why everything seems to be inverted and why the darkness of evil seems to triumph over the light of goodness, know this – it is planned by the hidden hand of the few, to be that way. But it doesn't have to be that way if people wake up, turn on the light of awareness and start seeing more clearly. Once we begin to change our beliefs, we also change the 'reality' that we experience. As the blinkers are taken off and the light of our consciousness starts to emerge so are we transformed and so eventually, is the world.

Although it is not within the remit of this book to reveal the finer detail of where it is planned for the human race to go (think George Orwell's 1984 and *then some*) there are several

Another Perspective

diligent and astute researchers whose books and websites expose more fully the agenda of those striving to diminish the light and plunge humanity into darkness and some of these are listed in the appendix. I strongly suggest that you carry out your own research and keep your mind open as you do, because the lies that we have been told about who and what we are and the manipulation to which we have been, and still are being subjected, will truly astound you. I can guarantee that you will never again see the world in the same way or be fooled as you have been up to this point. This realisation alone, is truly liberating and part of the same process of opening to the inner light of being that I have been revealing throughout. An enormous part of becoming 'enlightened' is not just becoming aware of wisdom and truth, it is also seeing clearly the enormous deception that has been imposed by the relatively few upon the majority.

Although there is a long way yet to go before enough of us wake up from our spiritual slumbers and smell the coffee, there has already been a sea change that has taken place in recent times. With the advent of the Internet and changes in the way that we communicate, more information is becoming both available and accessible and the word gets out much more quickly than it used to. They say that bad news travels fast, but so does other forms of information these days. The genie is emerging from the bottle and once it is out, it cannot be returned. Light is always greater than darkness and the darkness knows it. In the next chapter we will be looking at ways in which we can work with light, both internally and externally, to enhance our lives.

Light: The Divine Intelligence

7

Making Light Work

"There has never been a time on Earth like we see today. What we need are more ways to experience our interconnectedness - it is a precursor to deep love. So in this quickening light, with the dawn of each new day, let us look for love. Let us no longer struggle. Let us ever become who we most want to be. As we begin to be who we truly are, the world will be a better place."

John Denver (1943 - 1997)

These days you can buy 'How to....' books covering pretty well any subject – how to cook, how to fix your car, how to use a computer, how to meditate, you know the kind of thing. If you've ever seen the 'Dummies' range, you'll know that they cover pretty well everything you'll ever need to live a happy, contented life. There's even 'Physics of Light for Dummies', although I haven't used anything from it in compiling this book, honestly.

These types of reference books are great, especially if you genuinely know very little about a subject and want to get a grasp of the basics. One I found particularly interesting a few years back was 'Guitar for Dummies', but I must have been quite a dummy myself at the time because having bought it I never really devoted enough time to making it work for me. I guess you could say that the left-brained intellectual approach, designed to deliver methodical, factual information never struck a chord with me. I must confess to being the same with music in general. I love listening to it and I'm forever buying or downloading new music from my favourite bands, but when it comes to reading music (which I can to an extent) I quickly get bored. When I play either my guitar or keyboards, I always prefer to do so in an improvised way, even though that invariably leads to mistakes that irritate my lovely wife, who is able to sight read and sings in a choir, as well as in the past having played French horn to a high level. I guess my passion for music doesn't really extend far enough to make me want to spend hours practicing or learning the intricacies of composing or writing a score, although I do appreciate the brilliance of those who can. In my defence though, there are some wonderful musicians who can't read or write a note of music but who have composed some of the most amazing pieces. Amongst these are all four of The Beatles, Michael Jackson, virtuoso guitarist Tommy Emanuel, rock legends Jimi Hendrix, Eric Clapton and Van Halen, writers of Broadway musicals, including Irving Berlin, Lionel Bart and Anthony Newly, and most notably the high profile Hollywood film 'composer' Hans Zimmer who in his younger days had piano coaching only briefly because he disliked the discipline of formal lessons. Zimmer does the majority of his compositional work, sitting in front of his synthesizer with the aid of modern technology - although you'd never know it by listening to his majestic scores.

The fact is, composing and playing music is not about notation, its about feeling, creativity and inspiration. There has to be some application of course - developing a skill takes practice, but if the talent and ability is there, it only needs nurturing for it to flourish. Miles Davis, perhaps the most ambitious and celebrated jazz musician of our time, understood this better than most. He was actually enrolled in Music College in the 1940s to learn formal music theory, but dropped out because he felt it was holding back his creativity. Who can argue with that, as Davis went on to record some of the most genre-expanding music ever heard.

The aforementioned Eddie Van Halen, not known for holding back on his musical views (or anything else apparently) famously said:

"Obviously you have to have rhythm. If you have rhythm, then you can play anything you need. If you have rhythm and you love music, then play and play and play until you get to where you want to get. If you can pay the rent, great. If you can't, then you'd better be having fun."

How true that is, and it's the same with so many pursuits in life, especially spirituality. If you have the inner light within (and everyone has) you don't need to pursue an intellectual path in order to 'understand it' or 'activate it' - *you just have to allow it to express itself through you.*

Whilst doing some research for this book, I typed into my search engine 'light body' half expecting to be directed to a plethora of sites instructing me how to develop this, fix that or awaken the other – I wasn't disappointed. Well, actually I was, because there were lots of them, some worse than others and most requiring payment in exchange for their 'knowledge'. Now, whilst I'm not against study and acquiring information there is no substitute for walking the walk and 'being that, that you would become'. Reading about a subject, won't give you the *experience* of it. To get that, you have to immerse yourself within it, to be it. When I first began my own development as a medium around the age of seventeen, I began by reading book after book and watched and listened intently the work of established mediums. To me they seemed light years (deliberate use of the term) ahead of me and they gave me something to which to aspire but it was only when I began developing in earnest, sitting regularly in a group that my progress really intensified. In those early days, my main mentor had said to me more than once 'You will learn through your own lips'. He could just as well have said 'your own mind', but I didn't really appreciate the meaning of his words until much later in life when I began to have the sense that I was learning *directly*, from some invisible source, rather than indirectly through the words of others. In the context of light, it was as if I was able to look directly at the light, rather than that upon which it shone. I no longer needed to shield my eyes and view knowledge and wisdom through the pinhole projector of the mind, my own or anyone else's.

How had I arrived at this point? How do any of us reach this destination? Well, to begin is to recognise that there is no destination, no single point or source to be reached. As I have previously intimated - *we are already the inner light, all that ever was, is and ever shall be.* All that is necessary is to remove the blocks, take down the barriers, peel away the layers of ignorance built up by ages of conditioning and mind-created, linear-based beliefs and be that which you truly are. Any teacher or source of information offering you a course in 'Activating your light body' (sometimes the word 'activating' is replaced by 'awakening') or something similar should in my opinion be considered very carefully before any action is embarked upon. Nearly all of these contain a modicum of scientific fact, interspersed with pseudo-science, new age mumbo-jumbo and complete nonsense in varying amounts. If they offer you a certification or ask you to part with your hard earned cash, avoid them at all costs!

Don't misunderstand, I am not out to knock anyone and we all have something to offer in lesser or greater amounts, but I sometimes think that common sense goes out of the window when it comes to seeking spiritual enlightenment and the New Age movement in particular, has a lot to answer for. In many ways those that have a little knowledge, but believe they are on a mission to save the human race have muddied the waters. We live in the information age and one of the major protagonists, the Internet, whilst having innumerable benefits, is partly to blame for this state of affairs by facilitating the free-flow of information in ways that allow self-styled prophets to post absolute nonsense disguised as unquestionable truth. Although I am totally opposed to censorship and respect everyone's right to believe what they want to, I do wish that people would be more discerning before consuming garbage, disguised as fine dining. So let's keep things positive - aside from our own unique naturally occurring spiritual development and the accompanying increase in our personal inner light and associated energy curve, what can we do to enhance our progress? Quite a bit, as it happens.

As I see it

First off, it seems sensible to me that looking after our self should be a prime consideration. Unless we have entered into this frequency with a particular life plan that involves suffering as part of spiritual growth we should be experiencing reality through a holographic form that for the most part is healthy and functioning properly. Living in this vibratory state does take its toll on any biological system over 'time' but presupposing it hasn't done so to the extreme and that our consciousness is able to maintain its hold over it then light will be working its wonders throughout every level of being – holographic physical, mental and emotional. However, dependant upon the level of indoctrination to which we have been subjected by the state, society and our own family, our personal belief system will have developed in one of several ways each of which compromises to a degree the extent in which light and its attendant information flows through us. If for instance we hold strong religious views that demand adherence to strict practices and ways of living, these will impact directly upon our mental and emotional states and indirectly upon our 'physical' form. Continually focusing on one thing, particularly if it's something negative such as illness, has the same effect. Because we are *body-mind* there can be no real distinction between the various systems through which consciousness operates. They are only different rooms under the same roof and the entire house knows if one or more of its parts are compromised.

The power of the mind can never be understated. It acts as the prism through which the light of consciousness shines and in accordance with the unique configuration of each individual, influences expression in incalculable ways. So it makes sense to be mindful of our mind and be thoughtful of our thoughts. Better still if we can become the 'watcher' of both *(Fig 32)*.

Fig 32. You are the 'watcher' - the observer, observing the self.

When we become aware of being the observer of our *bodymind* and its associated thoughts, feelings and emotions, there exists the recognition that we need not be entirely bound by it. At these times the focus of our conscious attention shifts from the lens of the mind to the expansive state that lies outside of mind. Occasionally this occurs spontaneously, for no apparent reason and may last only momentarily, but with a little practice it can be induced and endure for a greater length of 'time'. Actually, the sense of time whilst in this altered state is vastly different and what may only be a few minutes of clock time may seem like much longer to the participant. I refer to this as being in 'global time' or *the eternal now* rather than linear time and it is often accompanied by a sense of deep stillness, silence and bliss.

Meditation

Those who engage in meditation, particularly TM would describe this as *Samadhi* but this is only a label. What matters most in my view is that in this state - which is a *natural state of being* and not *supernatural* as many believe, the inner light is radiating at its purest and most refined, unimpeded by the chatter of mind and the pull of emotions. A true sense of oneness is present here as is the all-pervading love of infinite consciousness and having experienced this, returning back to the heaviness of what most people experience as normal waking consciousness can be somewhat disappointing to say the least. But return we must, at least temporarily so that the play of our holographic life can continue to unfold as it is meant to.

Daily meditation then, is for some, a good starting point; both in maintaining health and establishing more defined pathways to our higher self and expanded states of being. I'm not going to recommend one style over another because all approaches have some validity. Certain approaches may be better suited to us than others, fit in with our lifestyle more easily and be more comfortable – rather like our choice of clothes or footwear, but without question all will require from us a certain amount of focus and attention. They say that there is 'no gain without pain' and whilst meditation should never be painful, it will mean some small adjustments to daily routine, which is never a bad thing. TM for example, only requires two, daily twenty-minute sessions – one in the morning and one in the evening, which is usually achievable for most people.

My understanding of meditation is that when engaged in regularly it can bring about untold benefits in health, resulting in lowered blood pressure, improved breathing and posture, a reduction in stress and anxiety, a balancing of the nervous and hormonal systems (reduced Cortisol) and an increase in brain coherence. Most importantly it allows the mind to be stilled, thoughts to subside and the ordinary thinking process transcended, facilitating a deeper connection with the *light of pure consciousness*. In modern scientific terms it helps establish a greater connection with the Unified Field.

The connection between meditation and light is frequently evident to both meditator and observer with the former being aware of seeing light, even with their eyes firmly shut and the latter witnessing it both surround and exude from the person meditating. Through a series of synchronous connections I discovered the work of Dr Frederick Lens (1950-1998) who became known as Rama and also Atmananda. Lenz was an American Buddha whose life was dedicated to teaching meditation and transmitting enlightenment. His initial experience of the deeper states of consciousness occurred when he was at the tender age of three (although I suspect he was an 'old soul'). He recalled sitting in his mother's garden as it dissolved into light and the material world disappeared (his focus of consciousness shifted, allowing him to decode a more refined reality). In this altered state he perceived the oneness of all things and it was only upon returning to his 'normal' waking state that duality returned. However, at the age of 30, after years of profound meditations and teaching others to meditate, Lenz realised an advanced state known as *nirvikalpa Samadhi* (said to be the highest state of Samadhi) that is ever-present. In this sense he became fully enlightened. What caught my eye when researching information about him was a paragraph I read on a website dedicated to his life and work that stated:

Rama effortlessly exuded the siddhas (powers) that are described in Buddhist and Hindu texts. For people attending Rama's talks, witnessing levitation, disappearance, beautiful colors and waves of golden light all around Rama and spreading through the hall or room - were commonplace. His attitude towards such phenomena was pure and humble - he said that if people saw the reality of the siddhas they would also believe that enlightenment was real.

So here again the pattern emerges – higher states of consciousness result in the ability to produce phenomena associated with light. This time, the individual in question had developed his abilities, or rather had acquired them through many years of devotional meditation and service to others. Lenz had as a child first attended Catholic schools and then, with his parents' divorce and his father's subsequent remarriage, attended Stamford public schools, neither of which could really be described as suitable environments in which to develop enlightened states. Perhaps though, they played a part in being the catalyst for the emergence of his desire for real knowledge and wisdom, as is often the case. Certainly for me, this was so – having rejected totally any orthodox teachings that were foisted upon me when I attended Sunday school as a boy scout because even at the age of eight I knew how self-limiting they were. What Lenz subsequently realised, as all enlightened beings do, is the nature of light and its fundamental role at the heart of everything. Here are a few of his quotes that I found particularly beautiful:

'Your mind is made up of light. We call it the dharmakaya, the clear light of reality. The transcendental eternal light is everywhere. It's the light of God or whatever you want to call it.'

'Life reorders you when you go into the clear light. Even the causal structure is liquefied. The clear light of reality, the dharmakaya, changes us into beings of light.'

'You perceive yourself as being solid and physical, limited to a body. You are made up of light, endless mind. The way you see life is a limited state of mind.'

'Within the universe there is a pure light. It is a light that is beyond all darkness. It does not give way to anything. It is the light of existence.'

'Enlightenment occurs when your mind merges with nirvana, with what Tibetans call the dharmakaya, the clear light of reality, which is the highest plane of transcendental wisdom and perfect understanding.'

'What you focus on, you become. If you spend and hour or two a day meditating and focusing on light, then you will eventually become light.'

These words reflect the consciousness of a man who had found many answers within – something that we all must do, sooner or later. Meditation was one of the tools that he employed because he recognised it as a means to an end, or should I say 'a means to infinity'. Most tellingly, in an elegantly worded quote Lenz affirms:

'As you go into light for longer and longer periods, as you progress in your meditation practice, you transform, you become illumined, you overcome all limitation, all sorrow and all pain. You learn not to be bound by desire and eventually you transcend death itself.'

What could be clearer than that?

Mindfulness

If meditation is a form of escaping mind, of going beyond mind, then mindfulness is the practice of maintaining a moment-by-moment awareness of our thoughts, feelings, bodily sensations and the surrounding environment. It is also an attitude of non-judgment, paying attention to our thoughts and feelings from the perspective of an impartial witness (watcher) and not believing them or taking them personally. In his famous book *The Power of Now* spiritual teacher Eckhart Tolle (1948 -) points out quite rightly, that most of us spend our 'time' focused in either the remembered 'past' or the imagined 'future' at the expense of what is happening *now*. The fact is; past, present and future exist simultaneously because *there is only now*. The way that humans (and all biological creatures in varying degrees) decode 'time' suggests otherwise but this is only an illusion created by the mind. Because of this, we often fail to notice what is happening around us and are absorbed in our thoughts of what did or didn't occur yesterday or might do so tomorrow. Being mindful of the present moment and being focused on what is happening now can prove extremely beneficial in lots of ways because there is more of an acceptance of our current reality when we let go of the tension caused by wanting things to be different, the tension of constantly wanting more - accepting instead, the present moment *as it is*.

Another benefit of living mindfully is that we become less judgmental. The aim is not to control, suppress or prevent thoughts from arising, simply to pay attention as they surface without judging or placing labels on them. Acting as the watcher also enables us to observe thoughts and emotions without being swept away by them.

There are two approaches to mindfulness that are equally valid and beneficial. The first is through meditation – also known as the formal approach and the second is through being fully aware in daily life - the informal approach. With the former, meditative techniques can be applied sometimes using controlled breathing or the use of mantras and sound, whichever is favoured. With the informal approach any routine activity can be made into a mindfulness practice provided *you give your full attention to it*. Again, the advantages to mental and physical wellbeing are well documented and the benefits far outweigh any issues that might be construed as negative.

Diet

Several years ago I decided to become vegetarian and flirted with the idea for several months. My wife was and still is a veggie and has been since her childhood. But I enjoyed the taste of meat and like most people had become used to bacon sandwiches, pork pies, chicken wings and a Sunday roast without really thinking about what I was eating. So I lapsed and returned to being a meat eater again, although it didn't sit quite right with me because of my strongly held spiritual views. Then, three years ago I saw a video clip taken inside an abattoir that rocked me to my core, such was the cruelty being inflicted by men upon innocent creatures. There was not one shred of decency, humility or compassion shown to animals being led to their death and worse still, they were being taunted and mocked by their executioners as if they were nothing. At that moment the decision to stop eating meat became easier for me and I made my mind up on the spot that I would never eat meat again.

Morally, I know that being a vegetarian is the right thing to do because all life is connected and not one creature is greater or lesser than another. What we give out, returns to us and our actions toward other life forms impacts upon us directly, whether we are aware of it or not. Firstly there is the moral issue and the simple fact that spiritual 'laws' operate across every spectrum, having a direct bearing on our progress, or lack of it. Understanding and awareness bring personal responsibility and acting from a position of knowledge carries more weight than when acting from ignorance.

Secondly, and what should also be of the utmost importance to every meat eater is the fact that every condemned creature is consumed by fear prior to its slaughter. Animals are extremely sensitive to smell, taste and feeling. They can sense fear and death well before they see the inside of the slaughterhouse and know what fate awaits them. They may well have travelled long distances in cramped conditions before arriving at their final destination and stress hormones released into their bodies has massive detrimental effects to human health. From the perspective of light alone, which is seldom if ever considered by medical science relative to meat consumption, there is a disturbance of energy as the information of fear is communicated throughout the DNA of the creature. Those choosing to consume meat subsequently ingest this genetic information and its message is communicated to their own DNA. Is it any wonder then that cancer and heart disease rates alone have increased amongst meat-eaters? They are unknowingly (for the most part) ingesting disease-causing information each time they eat a Big Mac. Red meat is obviously worse than white meat because of the amount of blood it contains, but neither is healthy because the programming doesn't only happen at the 'physical' level, but most significantly at a higher energetic level. Anyone with spiritual awareness who knows the facts but continues to eat meat is in effect agreeing to connect with the fear, pain and suffering of creatures bred for human consumption and even those who are ignorant still do so to a large extent, although in my view ignorance is no excuse for lack of empathy and compassion - we all have a conscience after all.

So what *should* we eat? I heard someone mention in all seriousness recently that 'if everything is alive, we shouldn't eat anything because we have to kill it'! I sort of see where they are coming from, but hold on a minute, wouldn't that have dire consequences for us all? All living forms on Earth need to take in energy and it isn't wrong for humans to obtain it through eating as long as the level of infinite consciousness within whatever we determine as 'food' has not reached the level of self awareness, where the thought processes of mind have been afforded the ability to function through the process of physical evolution. As a rule of thumb I'd say that if a species is self-determining and can think and reason beyond the basic instinctual level, displaying conscious choices, then we probably shouldn't consume it.

What we should aim for is a balanced, nutritional diet that meets the needs of our body whilst being respectful to the planet and all life upon it. Vegan and vegetarian diets fall within this remit and there is evidence that more and more people are embracing these options. Another beneficial system, although incorporating an optional amount of meat consumption is the ancient science of Ayurveda. This encompasses much more than just diet and incorporates a holistic approach to living designed to promote balance in mind and body. Some of the main Ayurvedic practices, relative to eating are basic common sense, yet surprisingly few people follow them:

- Always sit down to eat – never eat in front of your computer or TV or whilst driving
- Eat at a steady rate, not too quickly or too slowly
- Eat in a quiet, settled atmosphere and never when you are angry or upset
- Only eat when you are hungry
- Minimise consumption of raw foods - they are much harder to digest than cooked ones
- Include all six tastes at each meal (Sweet, Sour, Salty, Pungent, Bitter, Astringent)
- Drink hot water with ginger throughout the day

In addition, these daily practices are also highly beneficial:

- Some form of regular moderate exercise
- A daily oil massage with herbalised oil that balances your mind-body type
- Quiet meditation (morning and evening if possible)

Spiritual teacher and best selling author Deepak Chopra (1947 -) is so right when he says:

"The Ayurvedic approach is about aligning with the infinite organising power of nature rather than struggling or trying to force things to go your way. This principle is embodied by the 'Law of Least Effort.' When you observe nature, you will notice that grass doesn't try to grow; it just grows. Birds don't try to fly; they just fly. Flowers don't try to blossom; they just blossom. Nature functions with effortless ease, frictionlessly and spontaneously.

It is intuitive, holistic, non-linear, and nourishing. You will expend least effort when your actions are motivated by love, because nature is held together by the energy of love. When you chase after status, money, power, or accolades, you waste energy, but when your actions are motivated by love, your energy expands and accumulates. So take it easy and be guided by love."

Seems about right to me.

Our 'modern day' society is all about work, meeting deadlines, handling pressure and playing hard and fast. You often see people eating whilst standing or walking when they really should be sitting down and 'fast food' chains exploit this culture at the expense of human health because they know that many of us simply can't afford or don't make the time to eat our meals properly. Interestingly, we often use the phrase 'I can't afford the time' in relation to many things and this is so relevant because the price we pay for neglecting to eat properly (the same could be said of most of life's important activities) impacts on our wellbeing.

As I said earlier in the book, we obtain stored light energy when we eat healthily and relative to our spiritual growth and the development of our inner light it is important that our holographic 'physical' body reflects this. There is an ancient Biblical saying to be found in Corinthians 6:19 that states:

[19] 'Or do you not know that your body is a temple of the Holy Spirit who is in you, whom you have from God, and that you are not your own? [20] For you have been bought with a price: therefore glorify God in your body.'

Exercise

Applying the same common sense approach to exercise also helps our spiritual development. The *bodymind* vehicle through which we experience our version of reality functions better if it is looked after as nature intended. Muscles atrophy if not used regularly and vital organs become sluggish and eventually diseased if they are either abused or neglected.

Now, before I'm accused of discrimination against those who are less able bodied let me stress the point that developing the inner light and unfolding spirituality is not dependent upon ones physical status. Throughout history you will find numerous examples of highly spiritual individuals possessing of a body that was compromised in some way and it didn't hold them back. Indeed, it is often the case that the reverse happened and they were able to progress despite their physical challenges. Nature often has a way of compensating us in the most wonderful ways. As the sage said, *'He who is blind may see more than you or I.'* All that I am suggesting here is that if it is safe, comfortable and possible for regular exercise to be taken, then a sensible regime should be followed, even if it is just walking to the office each day or climbing a few flights of stairs instead of taking the lift. No one is suggesting running a marathon, although if you are fit enough, why not?

Usually whenever the word exercise is mentioned we think of the physical body. But the brain and the mind need exercising too. Meditation and mindfulness constitute a type of exercise but that's not what I'm referring to here. Most human thinking that is not from the level of the subconscious mind (a critical aspect nevertheless) is either repetitive or mundane - often the result of years of conditioning or else it is of the intellectual kind, which again is mostly conditioned. Seldom does anyone generate an *original* thought or idea. Original thinking comes from a deeper level of being – from the inner light. Often, it doesn't even require any mental process as we would understand it, just the ability of consciousness to focus upon the realm of all knowing and pluck the fruit of an idea from the tree of pure potentiality. When this occurs it's as if the thought just pops into the mind in a kind of eureka moment. I've had this happen to me on numerous occasions and every single time I instantly think 'Wow, where did that come from?'

Even if true originality seems spontaneous, it is only because the deeper nature of the individual seeks answers and *the intent is to know.* The person with no desire to know will remain unknowing, but where the inner light has begun to emerge, darkness of ignorance can remain no more. Exercising the mind by reading books that stretch the level of understanding and open up the possibility of new ideas and concepts is a start, as is engaging in meaningful dialogue with people that share similar interests. Contemplation - a fast vanishing ability in this day and age, is another rewarding pastime along with the observation of folk going about their daily business. I get great pleasure from sitting, drinking a coffee in the middle of a busy town square, just people watching and learning. Nature also has many lessons to offer us and there can be no finer teachers than trees, insects, birds, fish and the creatures of the field and we ignore them at our peril.

Bound up with exercise of the mind is exercise of the heart. We all know that the heart is a powerful muscle that pumps blood around the body and keeps us functioning in this frequency. We know that eating the wrong foods can clog up our arteries and lead to us having a heart attack or stroke. But what many of us seldom pay attention to is what our heart is *saying*. The heart is much more than just a pump - *it is the centre of our emotional mind.* For centuries, the heart has been considered the source of emotion, courage and wisdom and there is much truth in this. Research has discovered that there are more electrical connections routed *from* the heart to the brain, than vice-versa, influencing information processing, perceptions, emotions and health. Quite simply the heart is a highly complex, self-organised information processing centre with its own functional 'brain' communicating with the cranial brain through the nervous and hormonal systems and other internal pathways, influencing brain function and most of the body's major organs.

How often do we say 'I let my heart rule my head' or 'my heart is telling me.' Sometimes it even transpires that a person has apparently 'died of a broken heart' following a personal tragedy of some kind. This may be a turn of phrase, but it really is possible to 'die' in this way because intense emotions and powerful thoughts affect the electrical functioning of the heart and can be responsible for creating issues that ultimately have a direct bearing on health. When we 'think' from the heart or at least listen to what it is saying, we are again accessing the inner light. Heart-centred people nearly always display love, compassion, empathy, tolerance, kindness, gentleness and sensitivity in abundance

- as will any truly enlightened person.

What is known as 'Heart Intelligence' (also referred to as HQ) is a *higher level of awareness* that arises when we are able to integrate our physical, mental, emotional and spiritual intelligence together in a coherent way. I have gone on record before in stressing the importance of heart-centred thinking and devoted an entire chapter to this in *Transcognitive Spirituality*. I referred to the wonderful work of *The Institute of Heart Math* in the USA in exploring the potential of the human heart and in particular the torus-shaped electromagnetic field generated by it *(Fig 33)*.

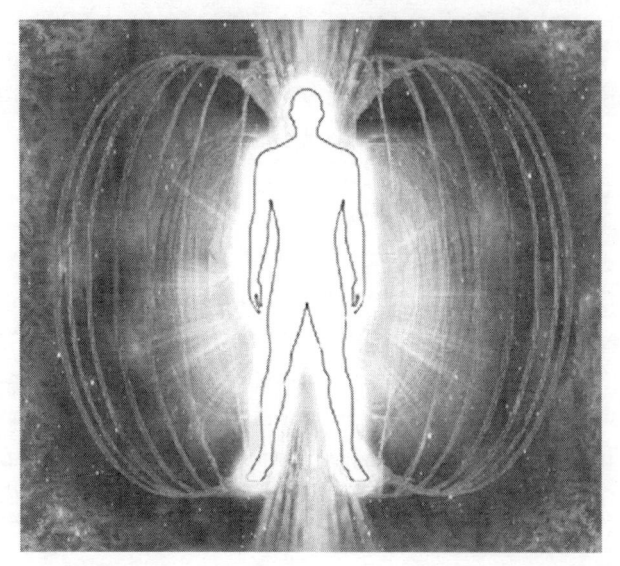

Fig.33 The torus-shaped electromagnetic field of the heart

I also wrote about what I term 'Coherent Loops' *(Fig 34)* which refer to the connection formed by coherent heart fields between humans and animals. The potential of these collective electromagnetic fields to introduce order into chaotic systems, restoring order and balance is immense. This too is the work of light. The heart's magnetic field is more than 5,000 times stronger than the same magnetic field generated by the brain so when you feel really happy or positive, this emotion not only creates a presence that can be detected by other people some distance from you but under the right conditions establishes a looped energetic connection with them, further amplifying the light. Imagine the light from a single torch or candle joining together with several others to create an even brighter light - truly powerful. But this is what the heart is capable of and even more so when we are aware of its potential and actively utilise this in a beneficial way. I recommend a visit to the Heart Math Institute website at *https://www.heartmath.org/gci/* where you will find information about *The Global Coherence Initiative,* an international effort that seeks to help activate the heart of humanity and promote peace, harmony and a shift in global consciousness.

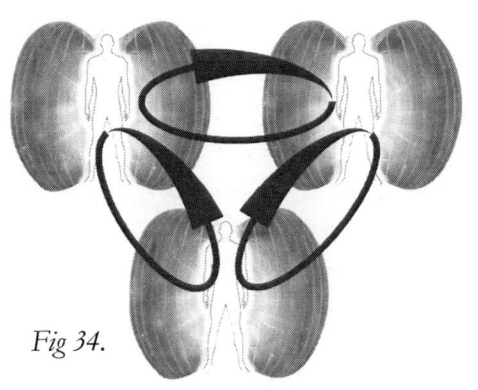

Fig 34.

A little more about the Pineal Gland

Earlier in the book (Chapter Two) we discussed the pineal gland and the role that it plays along with DNA as a receiver and transmitter of light and its associated information. This tiny gland acts as a bridge between dimensions - the holographic 'physical' and the metaphysical. It allows you to have mystical and lucid dream experiences that are part of your spiritual evolution. These are messages and guidance from your higher self and the pineal gland, along with DNA acts as an antenna to receive them. I've seen many sources describe methods in which we can 'activate' the pineal, as if it needed switching on. Nature has already done this of course, but what often occurs, sometimes with ageing and also through exposure to toxic substances such as fluoride is that calcification occurs, effectively limiting its ability to function and effectively closing it down. Assuming that you took my earlier advice and have viewed one or more of the websites that show you ways to decalcify the pineal I'm going to focus on how you can facilitate to a greater degree its operation relative to your inner light.

Whilst it's important to modify your diet and maintain a healthy body, nothing substitutes a daily spiritual practice that allows the healing to occur naturally, trains the mind to be still and prepares you to experience transcendence to higher states of consciousness. Activating your pineal gland, or spiritual 'eye' and detoxing it are two different things, though in essence, the detoxification allows you to remove any unwanted crystallised deposits and once the pineal gland is decalcified, you can begin practising various approaches that will help to 'switch it on' and increase natural production of your body's DMT (also referred to by some as the 'spirit molecule'). Because stress hormones like cortisol and adrenaline inhibit melatonin and serotonin, the sister hormones to DMT, practices such as meditation and yoga can have beneficial effects. Interestingly, we actually produce more melatonin when we are in complete darkness and this can help our inner light to shine. It is known that the ancient yogis and shamans would often meditate sitting in caves in order to induce higher levels of melatonin, and thus stimulate the pineal. Inversions (being upside down) are also particularly helpful because they increase blood flow to the brain. One of the most profound affects of yogic inversions is through what Indian gurus call the *'amrit,'* or 'nectar of Brahma'. The nectar they are referring to is the very same naturally occurring chemical DMT, which aids us in connecting to the more expansive metaphysical universe.

I am not qualified to advise on yogic methods or practices, of which there are many, but I would recommend anyone seriously interested in undertaking yoga as a means to enhance their spiritual development read one or more of the numerous expansive books available on the subject. Kriya Yoga, introduced by the aforementioned Paramahansa Yogananda (see Chapter Five), is a 'spiritually' based yoga, where an initiation takes place between the 'master' and his 'disciple'. Three basic principles are to be followed - self-discipline, introspection and devotion, through which the devotee eventually achieves a state of higher consciousness. These Kriya exercises should not be underestimated because they are not simple breathing exercises. The *'prana'* or *'life-force'* is controlled and channelled through the spinal passages with the spiritual energy or *'Kundalini'* being opened up, resulting in a heightened sense of balance and purity.

My own experience of 'activating' my pineal gland and subsequently the associated Kundalini energy came with many years of regular meditation in a small spiritual development group that resulted in the opening of my 'third eye'. This facilitated my development as a medium and allowed me to see clairvoyantly as well as to hear (clairaudiently) and sense (clairsentiently) those residing in the frequency that many term 'the spirit world.' More importantly it allowed me the privilege of being a recipient of information given through me whilst I was in an altered state (trance) as evidenced by the many demonstrations I have given over the years and the five books of spiritual philosophy that my wife and I have published. That period of sitting either in complete darkness or under subdued light has proved invaluable to me and to this day I retain the ability to penetrate into the subtle realms and higher frequencies in which lies great knowledge and wisdom. I must stress again that this was my way of development and in no way do I infer that I am superior or more advanced than anyone else. For others the 'journey' may be entirely different due to their unique personal circumstances and each must find their own way. To use the words of a higher intelligence *'what matters most is that the soul is touched.'*

Before concluding my thoughts about the pineal gland/third eye I will mention another process that I came across whilst doing my research for this book that I intuitively feel is relevant. As you are aware, in the opening chapter I recalled my time as a young boy, gazing through a pinhole apparatus that my father and I concocted to view a solar eclipse. Interestingly a 64-year-old mechanical engineer from India named Hira Ratan Manek (1937 -) also known as 'Hirachand', proposes a technique called 'Sun Gazing'. This is a process aimed at activating the pineal gland and it is claimed that he is now able to go without food for extremely long periods of time and still remain fit and healthy. Apparently Manek started disliking food in 1992 (as reported by the Hindustan Times) and in 1995 went on a pilgrimage to the Himalayas, stopping eating completely upon his return. According to his wife he obtains his sustenance directly from the Sun by gazing directly at it every evening for an hour, without batting an eyelid. Even NASA, the U.S space agency took an interest and invited him to spend time with them so that they could study the effects of surviving on water alone (which they later confirmed that he did for one hundred and thirty days). They even named this subsistence on water and solar energy after him: The HRM (Hira Ratan Manek) Phenomenon. NASA hope to use the knowledge gained to solve some of the problems encountered during longer space missions. Manek says he "eats through his eyes" in the evening, when the Sun's ultra-violet rays are least harmful and his wife claims the technique is totally scientific. She states:

"He has a special taste for Sun energy. He believes only 5 per cent of human brain cells are used by most people - the other 95 per cent can be activated through solar energy."

Putting aside the obvious dangers of staring directly at the Sun, I can see how the technique would work - why wouldn't it when the light of the Sun is the source of life on this planet? Manek's official website states:

'The method is used for curing all kinds of psychosomatic, mental and physical illnesses as well as increasing memory power and mental strength by using sunlight. One can get rid of any kind of psychological problems, and develop confidence to face any problem in life and can overcome any kind of fear including that of death within 3 months after starting to practice this method. As a result, one will be free from mental disturbances and fear, which will result in a perfect balance of mind. If one continues to apply the proper Sungazing practice for 6 months, they will be free from physical illnesses. Furthermore, after 9 months, one can eventually win a victory over hunger, which disappears by itself thereafter.

This is a straightforward yet effective method based on solar energy, which enables one to harmonize and recharge the body with life energy and also invoke the unlimited powers of the mind very easily. Additionally, it allows one to easily liberate from threefold sufferings of humanity such as mental illnesses, physical illnesses and spiritual ignorance.'

Could it possibly be that at some time in the distant past, humanity derived its 'food' directly from the Sun? If individuals such as Manek can go days or months without needing to eat and mystics and yogis can survive for even longer without food, is it that we were all once, of a similar constitution? Also - and I know it may be cynical of me to suggest this, it does raise the question of why there is a concerted campaign to keep people out of direct exposure to sunlight on the basis that it is apparently the prime cause of skin cancer when actually, with a bit more research and consideration it may be revealed that this is not the case. I assert that the Sun, and light in general have so much more to teach us than we are currently being lead to believe and perhaps we once instinctively knew much more than we do today. It is also interesting to note that on their 'Solar Heating' website there appears the following comment, which although not grammatically correct, mirrors much of my own thinking:

'HRM asserts that the rainbow is in the eye not in the sky. The seven colors of the Sun is only the reflection of what is in the eye. We can create a rainbow anytime we want – go to the garden, just observe below a source of flowing water as the Sun moves above. There you will see the rainbow. Eye can receive the entire spectrum of the sunlight. It's like having a glass window. Eye is the perfect instrument to receive all the colors of the rainbow. Since eyes are delicate parts of the body, we have to use them in such a way that they serve our purposes without getting damaged. Present day teachings and ideas such as 'don't look at the sunlight at all - you will damage your eyesight; never go out in the Sun as you will get cancer', are causing needless hysteria and paranoia. The more you are away from the nature, the more there is a cause for illness and you will automatically support global corporations. There are definite foolproof ways of getting the benefits of the nature without exposing ourselves to its adverse effects. It is also as intuitive as when the clouds gather we become gloomy. When we see the Sun, we feel energetic.'

It is far too easy to demonise things through the power of the mass media these days, especially anything that frees us from 'the system' and we need to awaken to this fact. Conveniently, that brings me one of the most critical points in this book.

The light of discernment

As the inner light grows ever stronger and exerts its influence upon every aspect of our being there begins to emerge an aura of knowingness that is reflected not just through our thoughts and actions but significantly in our judgments. We become more discerning and astute, more savvy and less easily fooled in any situation. When I look back at some of the things I did when I was younger, the mistakes I made and the bad decisions I took, I know that I would never make those errors again. Some would say that this is down to life experience, that the older we become, the wiser we are. There may be some element of truth in that but it is only part of the story and I have known many members of the older generation who are still as lacking in common sense and astuteness as they must have been when they were much younger. It is as if life has taught them nothing, or more accurately that they have simply chosen to ignore its teachings. When people say to me 'You are a wise soul' my reaction is always to think 'Relative to what, or to whom?'. I seldom respond to such praise because my personal criterion is always to remain humble.

What the inner light has a wonderful way of doing though, is to shed its radiance upon and within a given situation, illuminating it clearly for the watcher within me to see. Media hype, the official line or what the majority accepts without question as being the truth no longer engages me as it might have done. The mainstream 'system' that incorporates pretty much everything that is wrong with our sick society demands that we obey without question what it dictates to us. This is like asking infinite consciousness to obey the will of a microbe, except that the microbe probably has higher scruples.

It always amazes me how easily people will comply with their own enslavement and simply never question what they are told to do by those in 'authority.' Who gave them the authority in the first place, when the insane system that passes for 'democracy' only ever offers the illusion of choice? Opposites are always oppo*sames* in disguise.

I came across a quotation on the Internet taken from a story by Rabbi Nachman, titled *'The Seven Beggars'* that encapsulated beautifully the point that I am driving at here:

'Guard your thoughts very carefully, because thought can literally create a living thing. The higher the faculty, the further it can reach. You can kick something with your foot, but you can throw it even higher with your hand. With your voice you can reach even further, calling so someone far away. Hearing reaches further still, you can hear sounds like gunfire from a very great distance. Vision reaches even further: you can see things high in the sky. The higher the faculty, the farther it can reach. Highest of all is mind, which can ascend to the loftiest heights. You must therefore guard your mind and thoughts to the utmost.'

That is so very true. But I would add that the light of our

consciousness reaches still further, way beyond even mind. Armed with this knowledge ask yourself 'To what authority do I answer?' As an infant your immediate authority was your parents, then as a child it widened to include your school teachers. When you finished your education and started work as a young adult it became your employer. In the wider world you may have obeyed the authority of the military, but most certainly you would have been expected to obey the authority of law and of government. Yet always there was a greater authority, as there still is - *the authority of infinite consciousness, all that is, was and ever shall be*. Let this be your mentor and the authority to which you answer, not in any fearful, religious way, but in the light of knowing.

In all of your dealings be discerning. The 'physical' world that you decode into the reality that you experience is a harsh place. It is designed to trick you, snare you and hold you within its illusionary grip. But when your light shines fully, the mists disperse and your singular eye perceives clearly. When your consciousness penetrates deeply into the Unified Field there is a realisation that you are that, in every aspect. There is nothing else, only infinite love and infinite light.

8

Enlighten-*meant*

"Within each of us is a light, awake, encoded in the fibres of our existence. Divine ecstasy is the totality of this marvellous creation experienced in the hearts of humanity"

Osho (1931-1990)

You will have discovered by now that all that I have written about within these pages connects with your own spiritual evolution. Light, through each of its manifestations reaches out to us in a myriad of ways, some plainly obvious and simple but others unimaginably complex and hidden from our conscious mind. For the astronomer or astrophysicist observing the heavens, light is a phenomenon resulting from the known laws of the 'physical' realm. The same might be said of the quantum physicist, watching the strange behaviour of subatomic particles as they blink in and out of existence. But for those engaged upon the path towards enlightenment, light means something different - it is awakening from the sleep of ignorance to the remembrance of the boundless, infinite oneness of being - the true heaven.

Yet really, there is no difference. Light, whether visible or not, whether born out of the electromagnetic spectrum or integral to the unseen, inner spiritual spectrum, is still light. It is intelligent, benign, supreme and omnipresent. There is nowhere that light is not. It beckons us from the apparent darkness of our own ignorance drawing us like moths to its flame, only for us to melt into the bliss of its endless wisdom and eternal love.

Being a music buff, I listened recently to a CD by Seasick Steve in which he sings the line *'I started out with nothing and I still have most of it left'*. Well, alright Steve I get your drift. But I think what you could have said was, *'I started out with everything, forgot what I had, plunged into the darkness of ignorance and then found my way back to the light before realising that it was there all along!'* I guess though, that's why Mr. Seasick is a successful songwriter and musician and I'm not – his lyrics resonate with his followers and of course fit nicely in with the rhythm of the song. Mine though, might resonate deeply with you. As Steve's fans buy his recordings, you chose this book because the light contained within its pages had already connected with you at some deeper level long before you purchased it. Vibrationally there was a link between you and it – well, you and I really. Maybe it was the cover that attracted you, or the feel of the paper, or perhaps, as is more likely we share the same resonant field. Whatever brought you to our words, mine and those of the others that I've called upon to share with you this message; there was a reason for it. There are no accidents, only synchronicity and the light within me knows the light within you. There is a familiarity there and a deeper conversation taking place.

Author Jon Kabat-Zinn PhD (1944-) famously wrote *'Wherever you go, there you are.'* That is true in more than one sense. It is true from the level of the holographic 'physical' – who could argue against the fact that we are wherever our bodymind vehicle is at any one time? But, at another level we are everywhere, because *we are everything*. We are either infinite consciousness or we are not. From the level of form things appear unconnected, *but from the level of all that is, everything is connected* - so too with light. There is no darkness, only that which appears so. We have already suggested that 'dark matter' is only light that is not observable to us, and ignorance is the same - it is enlightenment unexpressed.

I remember giving a public demonstration of trance mediumship at Freshwater, a small town situated on the Isle of Wight, UK during which the intelligence speaking through me suggested that those who committed heinous crimes, although acting out of ignorance, were still in a strange way, seeking happiness. Some members of the audience found this deep teaching hard to comprehend and it was only later, upon reflection and with the help of my wife who has a remarkable way of recalling most of what is spoken through me, that I could see that he was right. At the deepest most fundamental level we are all seeking happiness and contentment from life - we are 'wired' to do so whether consciously aware of it or not. That a relative few of us resort to evil, violent acts in the misguided belief that we will reap whatever reward we might be seeking doesn't change the fact that essentially we desire something more, something better than what we have. An enlightened person would never act from the same low level of course, having transcended such states, but from the distorted perspective of ignorance, the essence of goodness can get twisted into something totally different. At the level of connectedness though, there is only the all-encompassing purity of light and love and so it has follow that each of us embraces this, but when the essential message that it holds becomes temporarily lost or distorted then human words and deeds deviate away from its centre and can appear as something else. I would emphasise that it should never be that we tolerate evil or ignore its consequences, only that we see into the heart of it and know that buried deeply, behind layer upon layer of ignorance exists the same divine spark that lies within us all.

Enlightenment is the prime message of light. It is *meant* that we should all reach our summit to realise the latent perfection that we each hold and ultimately this cannot be prevented, only delayed. That this process *is* delayed either by deliberate choice or due to other circumstances cannot deflect what is to be. It is not a race to see who gets past the finishing post first. The bliss of becoming and the emergence of the inner light is there for all to experience. It is beyond space and time, beyond relationships, beyond worlds and beyond words. The joy of remembering the light that we each are is indescribable. What some now have, *all will have*. The appearance of illumined souls, radiating light from every facet of their being, shall be our appearance too - yours and mine. The creator has no favourites and what is true for one, has to be for all, within the timelessness of eternity.

I will meet you there, that is for certain. But when we connect, it will not be as separate beings, but as one being. What I recognise within you and you within me, will be the same. We will see each other perfectly and know ourselves. We will join, but be already joined and the light that has never left us, that has always been there even in our darkest moments, will radiate forth more brightly than ever before.

I am reminded of the inspirational words of author and researcher Paul Brunton (1898-1981) who whilst traversing India in search of spiritual wisdom visited the ashram of the great sage Ramana Maharishi and became acutely aware of the essence of pure light and love radiated by the guru and who finally, after wrestling with his own intellect had the experience of genuine enlightenment that changed him forever:

'I find myself outside the rim of world consciousness. The planet which has so far harboured me, disappears. I am in the midst of an ocean of blazing light, The latter, I feel rather than think, is the primeval stuff out of which worlds are created, the first state of matter. It stretches away into untellable infinite space, incredibly alive.'

Despite the advice that I have imparted throughout this book, I cannot state, nor would I ever venture to, that I have attained enlightenment, just as no master would ever acknowledge their own transcendence. All I know is that there is much that I do not know and what understanding I have acquired has only revealed to me the great truth that there remains much yet to be understood. But like Paul Brunton and others I can say that I too have experienced the ocean of light around and within my being and have glimpsed the play of the formless upon the form and it has changed me.

There have been times in my life when I have effortlessly and unconsciously experienced moments of bliss, at first randomly and infrequent but more recently on a daily basis and for longer periods. I cannot categorically state that meditation has invited these exquisite moments to descend upon me or that any spiritual knowledge and wisdom that I may have been fortunate enough to have attained are the catalyst. Neither can I claim to have consciously activated any gland or sacred centre within myself or to have vigorously pursued any ritual designed to awaken my higher spiritual being other than meditation. But perhaps through my sincere efforts to progress, through trying to be kind and selfless and through serving others, despite my many shortcomings, the inner light has found a way of lifting me above this mundane world, if only temporarily.

I enjoy reading books, but I am not academic and intellectual conversation bores me rigid. Even though the love I have for my family is immense and I enjoy their company, I am equally happy being solitary almost to the point of unintentionally alienating those around me. I also revel in silence, enjoying the peace that it imparts, so I guess you could say I'm something of a loner.

When I do 'meditate' it is usually only for twenty to thirty minutes at a time. I have always been comfortable with simply sitting in darkness, perhaps because I know that after a few moments, my vision will begin to catch glimpses of light dancing before them and with my eyes tightly closed I often see coloured lights coming and going and white light exploding all around me. I feel the warmth and closeness of a presence other than my own, one that is strangely familiar and this is sometimes accompanied by the most beautiful perfume like the nectar of the Gods. After a short while my vision extends and widens, allowing me to 'see' without bodily movement, all around. Then my consciousness seems to extend until the self that I am familiar with is no more. The 'me' that I know dissolves, having been replaced by a boundless sense of oneness with all that is and there is no conception of where this begins or ends. Yet enfolded somewhere within this blanket of darkness that is light, silence that speaks and stillness that moves, is my awareness.

There is revealed the mystery of creation, the dance of particles as they explode seemingly from nothingness into everything. There is the fusion of elements, the birth of stars and of worlds and the emergence of individual consciousness, at first rudimentary and then growing in self-awareness and power. Everywhere there is movement and intelligence. Life in all its rich diversity and abundance explodes as forms of every description arise and multiply, evolving, developing, learning and expanding.

Darkness dissolves to be replaced by abundant, all-encompassing light and contained within each photon, the energy of life itself, indestructible and incorruptible, now radiating forth from stars and atoms in equal measure as it reaches deeply into the DNA of all living form. At this level a great and vital conversation begins, one that appears indiscernible and yet is absolutely essential. Here, light performs its greatest work, like some secret messiah converting all unto its doctrine of love, unseen by the masses. Man professes to have unravelled the language of DNA, but the language of light remains mysterious to him and he witnesses only its effects. Here though, 'I' experience an understanding of its very nature and know that infinite consciousness and pure light is one and the same.

All of this occurs in what seems like hours, yet only minutes of clock time have passed. But that is the nature of the illusion and a glimpse into infinity through an altered state can allow the deliverance of large amounts information instantaneously, proving that we are parts of an indivisible ocean of consciousness.

Knowing the light, working with it, allowing it to reveal its innermost power is within the capability of each of us. Despite what may seem insurmountable problems or inherent difficulties, the potential for enlightenment resides at the heart of every single individual. It may seem that some of us are born with disadvantages that prevent us from doing what we would wish to and achieving all that we would like in life but this is true only to an extent. Metaphysical laws operate in conjunction with individual spiritual development that cannot be transgressed and yet at the right time, when circumstances allow, the pathway emerges and the way becomes known.

I found a true story about one of the founder members of Alcoholics Anonymous, a chap named Bill Wilson who was almost at death's door, having collapsed following a drinking binge (presumably before creating the organisation). As he lay in his hospital bed he experienced something remarkable. I quote:

"Suddenly, my room blazed with an indescribably white light. I was seized with an ecstasy beyond description……A wind, not of air, but of spirit [blew through me]. In great, clean strength it blew right through me. Then came the blazing thought, "You are a free man."…….A great peace stole over me and……I became acutely conscious of a presence which seemed like a veritable sea of living spirit. I lay on the shores of a new world."

I really don't think that what occurred to Bill that day was due to his drinking, but *despite* it and he was almost certainly at the point of being spiritually and energetically ready for it to happen. This scenario often plays out when one is close to death, sometimes through an out of body experience or a deep insight into one's own human condition. What does this tell us? It tells us in the most profound way that when we are ready to make a breakthrough and transcend to another 'level' of awareness it will happen. As much as we strive to develop and unfold our spirituality through self-discipline, devotion, service and intent, sometimes it is deep suffering that acts as the key to unlock the door to enlightenment. In truth, it matters not what the catalyst is or when the change happens, only that it does. When the inner light comes calling, it has a way of communicating with each of us in its own unique way. For some, it enters through pain, for others it is by virtue of their spiritual endeavour across several 'lifetimes' - *but call it does.*

The light belongs to us all, because *we are the light*. Collectively, we may have forgotten this, but fortunately for us the light hasn't. So open the door of your heart and the windows of your mind and let this visitor in. Sooner or later it will transform you and it *will* enlighten you. It has to, because it is *meant* to be.

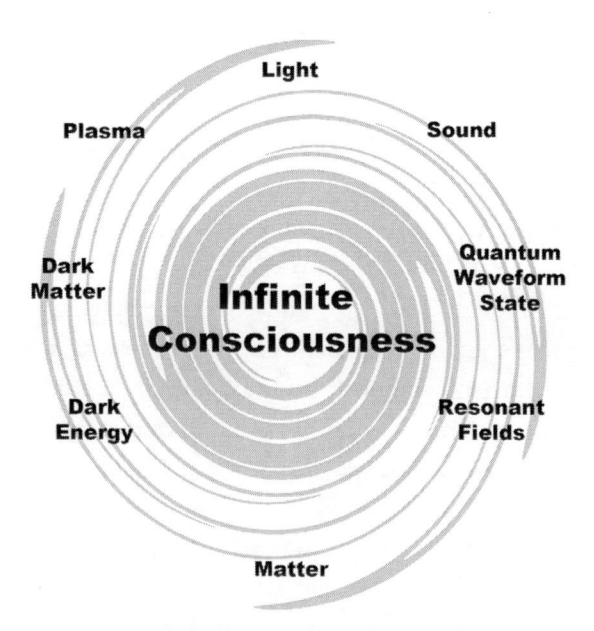

Light
Plasma   *Sound*
Dark Matter   **Infinite Consciousness**   *Quantum Waveform State*
Dark Energy   *Resonant Fields*
Matter

I am everywhere and everything,
In the seen and the unseen,
In the known and the unknown.
I am the divine intelligence,
that sings to your cells and plays upon the strings of your heart.
I am the formless, dancing within the form,
the infinite source of all life.
I am light and I am *you*.

Postscript

Over forty-years ago I received information that in my current earth lifetime I would witness many advances in the use of both light and sound, particularly in the field of medicine. Around the same time I was also given a warning about the future and was told that there would arise a 'situation that would affect everyone on the planet that would be difficult to avoid'. The former has most certainly come to pass with light being used by virtually everyone in the 'civilised' world on a daily basis across a wide variety of devices - everything from a TV remote to fibre optic broadband. Lasers are used to correct short-sightedness and remove cataracts as well as unwanted birth marks and skin blemishes and today they are routinely employed in most medical disciplines, including dermatology, dentistry, cardiology and neurosurgery, preventing the need for invasive surgery. They have also been used for over 20 years in treating some cancers. Photodynamic therapy uses a drug to make cancer cells more sensitive to light. When a laser is directed at the area of the cancer, the drug is activated and the cancer cells are destroyed.

The latter is also unfortunately coming to pass and is through no coincidence linked to light. As I mentioned in the final chapter of this book, it cannot have escaped your notice that our skies are sprayed almost daily with a cocktail of chemicals that spread out and very quickly block the sun, preventing it from delivering untold benefits to all life on the planet. This is happening right now, across the globe. Governments trot out lame excuses about 'climate change' and 'geo engineering' none of which wash with me and unless we collectively address this issue and who is really behind it, rather than pretending it isn't happening, the better. Below are some useful website links to those who are aware of this issue and I would urge those of you who care about the future of your children and grandchildren to check them out before deciding what, if anything, you can do. Natural sunlight and normal skies (when was the last time you ever witnessed a normal sky?) are our natural heritage and we cannot, must not allow anyone or anything to take these away from us.

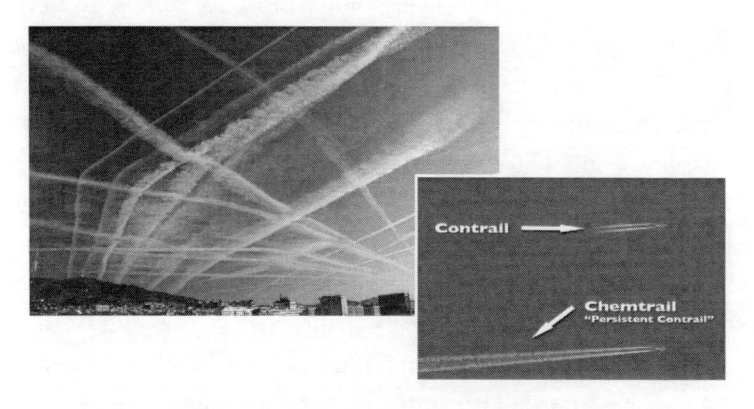

http://www.chemtrailsprojectuk.com/
http://www.chemtrailcentral.com/
http://educate-yourself.org/ct/
http://www.uk-skywatch.co.uk/
https://www.davidicke.com/

Bibliography

1. **Kelly,** R., *The Human Antenna:* Elite Books 2007, *The Human Hologram:* Elite Books 2011
2. **Kilner,** W., *The Human Atmosphere:* Publisher unknown, 1911
3. **Bagnall,** O., *The Orgin & Properties of the Human Aura:* Red Wheel/Weiser, 1975
4. **Burr,** H.S., *The Nature of Man & the Meaning of Existence:* Thomas, 1962, *Blueprint for Immortality:* Daniel & Co.,1972
5. **Horgan,** J., *The End of Science:* Basic Books (new edition), 2015
6. **Lazlo,** E., *Science & the Akashic Field:* Inner Traditions, 2007
7. **Huxley,** A., *The Perennial Philosophy* Haper Perennial (new edition), 2009
8. **Frankl,** V., *Mans Search for Meaning:* Rider (new edition), 2004
9. **Maltz,** M., *Psycho Cybernetics:* Perigee Books; (Upd Exp edition) 2015
10. **Yogananda,** P., *Autobiography of a Yogi:* Yogoda Satsanga Society of India, 2013
11. **Icke,** D., *Phantom Self:* David Icke Books, 2016
12. **Tolle,** E., *The Power Of Now:* New World Library, 1999
13. **Goodwin,** R., *Transcognitive Spirituality:* R.A.Associates, 2013

Online Sources

http://ancienthistory.about.com/od/sungodsgoddesses/a/070809sungods.htm

http://reviewofreligions.org/2306/ancient-sun-worship/

https://www.quora.com/What-is-beyond-the-electromagnetic-spectrum

https://en.wikipedia.org/wiki/Visible_spectrum

http://www.viewzone.com/dnax.html

https://en.wikiversity.org/wiki/Human_vision_and_function/Part_1:_How_the_eye_works/1.3_Light_stimulus_and_the_eye

http://www.universetoday.com/74027/what-are-photons/

http://math.ucr.edu/home/baez/physics/Relativity/SpeedOfLight/speed_of_light.html

http://coolcosmos.ipac.caltech.edu/cosmic_classroom/cosmic_reference/redshift.html

https://www.spacetelescope.org/about/history/the_man_behind_the_name/

http://www.physlink.com/Education/AskExperts/ae384.cfm

http://www.aps.org/programs/outreach/history/historicsites/penziaswilson.cfm

http://hubblesite.org/hubble_discoveries/dark_energy/de-what_is_dark_energy.php

http://curious.astro.cornell.edu/about-us/108-the-universe/cosmology-and-the-big-bang/dark-matter/658-what-s-the-difference-between-dark-matter-and-dark-energy-intermediate

http://planetfacts.org/how-does-the-sun-produce-energy/

http://www.plasmacosmology.net/

http://sunearthday.nasa.gov/2007/locations/ttt_heliosphere_57.php

http://www.transpersonal.de/mbischof/englisch/webbookeng.htm

http://www.techworld.com/news/personal-tech/dna-molecules-can-teleport-nobel-prize-winner-claims-3256631/

http://www.livescience.com/50678-visible-light.html

http://ngm.nationalgeographic.com/2015/03/luminous-life/judson-text

http://www.andor.com/learning-academy/what-is-light-an-overview-of-the-properties-of-light

Online Sources

http://www.fromquarkstoquasars.com/is-music-in-our-dna/

http://www.electricuniverse.info/Electric_Sun_theory

http://news.sciencemag.org/2006/01/and-galaxy-said-let-there-be-light

http://www.nikon.com/about/feelnikon/light/chap01/sec01.htm

https://en.wikipedia.org/wiki/Dark_matter

http://www.globalhealingcenter.com/natural-health/everything-you-wanted-to-know-about-the-pineal-gland/

http://www.ncbi.nlm.nih.gov/pubmed/15589268

https://www.newscientist.com/article/dn4174-plasma-blobs-hint-at-new-form-of-life/

http://beforeitsnews.com/space/2014/09/lightning-storm-formation-frequency-and-severity-linked-by-researchers-to-high-speed-solar-winds-2483282.html

http://www.gardeningknowhow.com/plant-problems/environmental/how-light-affects-the-growth-of-a-plant-problems-with-too-little-light.htm

http://health.howstuffworks.com/pregnancy-and-parenting/pregnancy/fetal-development/how-a-fetus-grows.htm

http://rationalwiki.org/wiki/When_does_life_begin%3F

http://www.i-sis.org.uk/TheRealBioinformaticsRevolution.php

http://themindunleashed.org/2014/06/electromagnetic-field-around-every-person-becomes-depleted-unhealthiness.html

http://lieske.com/hef.htm

http://www.wrf.org/men-women-medicine/dr-harold-s-burr.php

http://www.context.org/iclib/ic06/gilman2/

http://ascensionglossary.com/index.php/Blueprint

http://ervinlaszlo.com/index.php/publications/articles/96-are-you-available-augmenting-access-to-the-akasha-dimension

http://www.evolvingbeings.com/essay/the-physics-of-consciousness-the-zero-point-field-pineal-gland-and-out-of-body-experience

Online Sources

https://en.wikipedia.org/wiki/Sonoluminescence
http://enthea.org/library/brilliant-disguise-light-matter-and-the-zero-point-field/
http://www.theorderoftime.com/science/sciences/articles/spectrumofcons.html
https://en.wikipedia.org/wiki/Viktor_Frankl
http://www.huffingtonpost.com/2014/09/16/solid-light-created_n_5824268.html
http://www.angelfire.com/yt/Yukteswar/
http://www.kriya.org/about__guru.php?id=2
http://raybrownhealing.com/
http://abcnews.go.com/Nightline/gurus-mystics-india-performing-miracles-magic-tricks/story?id=11929407
http://www.ananda.org/autobiography/#chap30
http://www.space.com/30124-is-our-universe-a-fake.html
http://www.ramameditationsociety.org/rama-dr-frederick-lenz
http://www.chopra.com/ccl/what-is-ayurveda
https://www.heartmath.org/resources/downloads/science-of-the-heart/
http://www.spacedaily.com/news/food-03d.html
http://solarhealing.com/
http://www.stevenmtaylor.com/essays/spiritual-alchemy/#4

These links are provided for informational purposes only and do not constitute endorsement of any products or services provided by these websites. The links are subject to change, expire, or be redirected without any notice.

About the author

Robert Goodwin was born in England in 1954. He grew up in his native city of Birmingham and pursued an interest in mediumship during his late teens when he became aware of the latent ability he had to connect to the afterlife. After sitting in a small group for several years to develop his gift, he began demonstrating publicly in 1979.

Although a fully qualified Hypnotherapist and NLP practitioner, Robert is best known for his work as a trance medium and along with his wife Amanda, frequently demonstrates throughout the UK and Europe. Together they have published several books of spiritual philosophy and their work has also been featured in Psychic News, the UK's foremost Spiritualist publication. Robert has also appeared on local radio and his popular website features recordings of his interviews and inspirational videos, including extracts from public demonstrations of trance.

www.whitefeather.org.uk

Now read the companion book to
Light: The Divine Intelligence

Transcognitive Spirituality
....bringing together spirituality and science like never before!
ONLY £12 plus p+p
Available online now from:
www.whitefeather.org.uk